22 ⁸
d₄
6/48

Princeton
Problems in Physics
with Solutions

D0060856

Princeton
Problems in Physics
with Solutions

Nathan Newbury
Michael Newman
John Ruhl
Suzanne Staggs
Stephen Thorsett

Princeton University Press
Princeton, New Jersey

Copyright © 1991 by Princeton University Press
Published by Princeton University Press, 41 William Street,
Princeton, New Jersey 08540
In the United Kingdom: Princeton University Press, Oxford
ALL RIGHTS RESERVED

Princeton University Press books are printed on
acid-free paper, and meet the guidelines for perma-
nence and durability of the Committee on Production Guidelines for Book Longevity of the Council
on Library Resources

Printed in the United States of America

(Pbk.) 10 9 8 7 6 5 4 3 2

Library of Congress Cataloging-in-Publication Data

Princeton problems in physics, with solutions / Nathan Newbury . . . [et
al.].
    p.   cm.
Includes bibliographical references and index.
ISBN 0-691-02449-9
1. Physics–Problems, exercises, etc.   I. Newbury, Nathan, 1964–   .

QC32.P75   1991   530'.076—dc20   90-47703

To our parents

Contents

Preface

No one expects a guitarist to learn to play by going to concerts in Central Park or by spending hours reading transcriptions of Jimi Hendrix solos. Guitarists practice. Guitarists play the guitar until their fingertips are calloused. Similarly, physicists solve problems. And hopefully, physicists practice solving problems until doing so seems easy. (Then they find harder problems.)

This book provides a collection of challenging problems for physics students at a range of levels. Some problems, particularly those in the first four chapters, require only an undergraduate physics background. The later chapters cover material that is frequently not encountered until graduate school. Don't be discouraged if some (or most!) of the problems are, in fact, challenging. That's the idea.

However, this is not only a problem book. We also provide complete solutions for each problem. These solutions assume a certain amount of familiarity with the topics, but are not written for experts. For this book to be of maximum benefit, of course, the solutions should be considered a last resort. *Try to solve the problems before looking at the solutions!*

The problems presented here were culled from general examinations written at Princeton University in the last ten years. All physics graduate students at Princeton must pass the generals examination before beginning their theses. The examination is split into two parts: "Prelims," usually taken in the first year of study, and "Generals," typically taken in the second year.

The preliminary examination (Prelims) covers the subjects usually studied at the undergraduate level. It is a six hour exam, taken in two days. Prelims has four sections: mechanics, electricity and magnetism, nonrelativistic quantum mechanics, and thermodynamics and statistical mechanics. The second examination (Generals) covers more advanced topics. The written part of this exam takes place in three three-hour sessions and comprises five sections: condensed matter physics, general relativity and astrophysics, nuclear physics, elementary particle physics, and atomic and "general" physics. Many of the solutions given here are *much* more detailed and complete than would be expected during the general examination. (The passing mark at Princeton is 50%, and students are allowed to choose from several questions in each field.)

We have divided this book into the nine obvious sections suggested by the format of the exams. There are ten problems in each of the sections. We do not make any guarantees that we have provided a random sampling of problems. The problems chosen are those that we found interesting, informative, and well-posed. We have also tried to avoid archetypal problems whose answers have been printed in numerous other books. In several places we have noted useful references. Although each solution has been checked and rechecked, occasional errors may have slipped through. We welcome comments, criticism, and corrections.

A group of five authors can run up an amazing list of debts in the writing of a single book. Our first, and most obvious, is to the many members of the Princeton University Physics Department who have written original, interesting, and instructive problems for the preliminary and general examinations. Their work forms the foundation of this book. We also thank those professors who read chapters of our draft and made numerous useful suggestions: Paul Chaiken, Aksel Hallin, Will Happer, Peter Meyers, Phuan Ong, Jim Peebles, Jeff Peterson, Sam Treiman, and Neil Turok. Early encouragement from Jim Peebles and Jeff Peterson was invaluable, as were the computer facilities of Joe Taylor and Mark Dragovan and the support of Dave Wilkinson. Our most heartfelt thanks, however, goes to all of our fellow graduate students, far too numerous to name, especially the members of our own prelims and generals study groups, who made this book possible.

Nathan Newbury
Michael Newman
John Ruhl
Suzanne Staggs
Stephen Thorsett

Princeton, August 7, 1990

Part I

Problems

Chapter 1

Mechanics

Problem 1.1. A Wham-O Super-Ball is a hard spherical ball of radius a. The bounces of a Super-Ball on a surface with friction are essentially elastic and non-slip at the point of contact. How should you throw a Super-Ball if you want it to bounce back and forth as shown in Figure 1.1? (Super-Ball is a registered trademark of Wham-O Corporation, San Gabriel, California.)

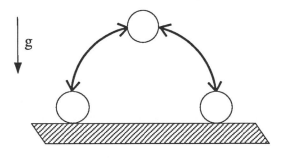

Figure 1.1.

Problem 1.2. Suppose a spacecraft of mass m_0 and cross-sectional

3

area A is coasting with velocity v_0 when it encounters a stationary dust cloud of density ρ. Solve for the subsequent motion of the spacecraft assuming that the dust sticks to its surface and that A is constant over time.

Problem 1.3.   The science fiction writer R. A. Heinlein describes a "skyhook" satellite that consists of a long rope placed in orbit at the equator, aligned along a radius from the center of the earth, and moving so that the rope appears suspended in space above a fixed point on the equator (Figure 1.2). The bottom of the rope hangs free just

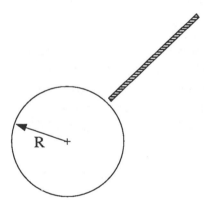

Figure 1.2.

above the surface of the earth (radius R). Assuming that the rope has uniform mass per unit length (and that the rope is strong enough to resist breaking!), find the length of the rope.

Problem 1.4.   Three identical objects of mass m are connected by springs of spring constant k, as shown in Figure 1.3. The motion is

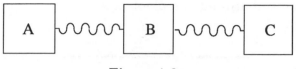

Figure 1.3.

confined to one dimension. At $t = 0$, the masses are at rest at their equilibrium positions. Mass A is then subjected to an external driving force,

$$F(t) = f \cos \omega t, \quad \text{for } t > 0. \tag{1.1}$$

Calculate the motion of mass C.

Problem 1.5. A uniform density ball rolls without slipping and without rolling friction on a turntable rotating in the horizontal plane with angular velocity Ω (Figure 1.4). The ball moves in a circle of radius

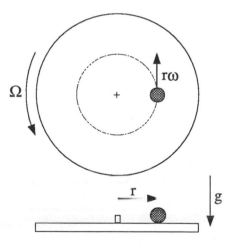

Figure 1.4.

r centered on the pivot of the turntable. Find the angular velocity ω

of motion of the ball around the pivot.

Problem 1.6.   A blob of putty of mass m falls from height h onto a massless platform which is supported by a spring of constant k. A dashpot provides damping. The relaxation time of the putty is short compared to that of putty-plus-platform: the putty instantaneously hits and sticks.

a) Sketch the displacement of the platform as a function of time, under the given initial conditions, when the platform with putty attached is critically damped.

b) Determine the amount of damping such that, under the given initial conditions, the platform settles to its final position the most rapidly without overshoot.

Problem 1.7.   A mass m slides on a horizontal frictionless track. It is connected to a spring fastened to a wall. Initially, the amplitude of the oscillations is A_1 and the spring constant is k_1. The spring constant then decreases adiabatically at a constant rate until the value k_2 is reached. (For example, suppose the spring is being dissolved by nitric acid.) What is the new amplitude?

Problem 1.8.   A soap film is stretched between two coaxial circular rings of equal radius R. The distance between the rings is d. You may ignore gravity. Find the shape of the soap film.

Problem 1.9. A bead of mass m slides without friction on a circular loop of radius a. The loop lies in a vertical plane and rotates about a vertical diameter with constant angular velocity ω (Figure 1.5).

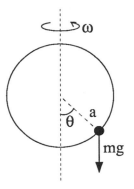

Figure 1.5.

a) For angular velocity ω greater than some critical angular velocity ω_c, the bead can undergo small oscillations about some stable equilibrium point θ_0. Find ω_c and $\theta_0(\omega)$.

b) Obtain the equations of motion for the small oscillations about θ_0 as a function of ω and find the period of the oscillations.

Problem 1.10. If the solar system were immersed in a uniformly dense spherical cloud of weakly-interacting massive particles (WIMPs), then objects in the solar system would experience gravitational forces from both the sun and the cloud of WIMPs such that

$$F_r = -\frac{k}{r^2} - br. \tag{1.2}$$

Assume that the extra force due to the WIMPs is very small (i.e., $b \ll k/r^3$).

a) Find the frequency of radial oscillations for a nearly circular orbit and the rate of precession of the perihelion of this orbit.

b) Describe the shapes of the orbits when r is large enough so that $F_r \approx -br$.

Chapter 2

Electricity & Magnetism

Problem 2.1.   A conductor at potential $V = 0$ has the shape of an infinite plane except for a hemispherical bulge of radius a (Figure 2.1). A charge q is placed above the center of the bulge, a distance p from

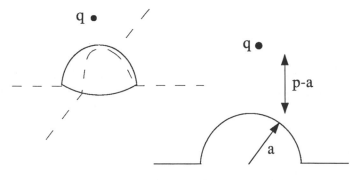

Figure 2.1.

the plane (or $p - a$ from the top of the bulge). What is the force on the charge?

Problem 2.2. A "tenuous plasma" consists of free electric charges of mass m and charge e. There are n charges per unit volume. Assume that the density is uniform and that interactions between the charges may be neglected. Electromagnetic plane waves (frequency w, wave number k) are incident on the plasma.

a) Find the conductivity σ as a function of w.

b) Find the dispersion relation — i.e., find the relation between k and w.

c) Find the index of refraction as a function of w. The plasma frequency is defined by $w_p^2 \equiv 4\pi n e^2/m$, if e is expressed in esu. What happens if $w < w_p$?

d) Now suppose there is an external magnetic field $\mathbf{B}_0$. Consider plane waves traveling parallel to $\mathbf{B}_0$. Show that the index of refraction is different for right- and left-circularly polarized waves. (Assume that the magnetic field of the traveling wave is negligible compared to $\mathbf{B}_0$.)

Problem 2.3. A cylindrical resistor (Figure 2.2) has radius b, length L, and conductivity σ_1. At the center of the resistor is a defect consisting of a small sphere of radius a inside which the conductivity is σ_2. The input and output currents are distributed uniformly across the flat ends of the resistor.

a) What is the resistance of the resistor if $\sigma_1 = \sigma_2$?

b) *Estimate* the relative change in the resistance to first order in $\sigma_1 - \sigma_2$ if $\sigma_1 \neq \sigma_2$. (Make any assumptions needed to simplify your method of estimation.)

c) Suppose $L \to \infty$ and $b \to \infty$, but a uniform current density j_0 continues to flow across the ends of the resistor. Calculate the current

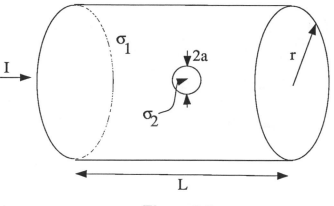

Figure 2.2.

density inside the spherical defect.

Problem 2.4. A thin, straight, conducting wire is centered on the origin, oriented along the $\hat{z}$-axis, and carries a current $\mathbf{I} = \hat{z} I_0 \cos \omega_0 t$ everywhere along its length l. Define $\lambda_0 \equiv 2\pi c/\omega_0$.

a) What is the electric dipole moment of the wire?

b) What are the scalar and vector potentials everywhere outside the source region (at a distance $r \gg l$)? State your choice of gauge and make no assumption about the size of λ_0.

c) Consider the potentials in the regime $r \gg l \gg \lambda_0$. Qualitatively describe the radiation pattern and compare it to the standard dipole case, where $r \gg \lambda_0 \gg l$.

Problem 2.5. As shown in Figure 2.3, a wheel consisting of a large number of thin conducting spokes is free to rotate about an axle. A brush always makes electrical contact with one spoke at a time at the

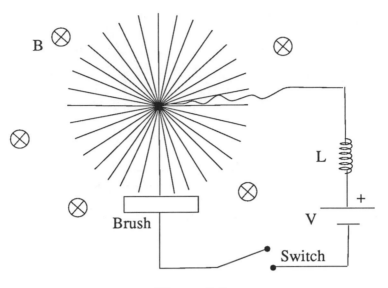

Figure 2.3.

bottom of the wheel. A battery with voltage V feeds current through an inductor, into the axle, through a spoke, to the brush. A permanent magnet provides a uniform magnetic field $\mathbf{B}$ into the plane of the paper. At time $t = 0$ the switch is closed, allowing current to flow. The radius and moment of inertia of the wheel are R and J repectively. The total inductance of the current path is L, and the wheel is initially at rest. Neglecting friction and resistivity, calculate the battery current and the angular velocity of the wheel as functions of time.

Problem 2.6.   A right-circular cylinder of radius R, length L, and uniform mass density ρ has a uniform magnetization $\mathbf{M}$ parallel to its axis. If it is placed below an infinitely-permeable flat surface, it is found to stick for some lengths $L \gg R$. What is the maximum length L such that the magnetic force prevents the cylinder from falling due to gravity?

Problem 2.7.   Consider an infinitely long transmission line which consists of lumped circuit elements as shown in Figure 2.4. Find the dispersion relation (ω versus λ) for periodic waves traveling down this line. What is the cut-off frequency?

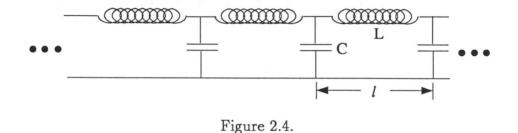

Figure 2.4.

Problem 2.8.   In 1890, O. Wiener carried out an experiment which may be said to have photographed electromagnetic waves (Figure 2.5).

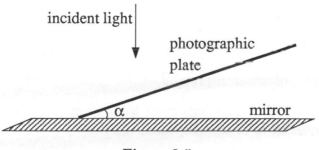

Figure 2.5.

a) A plane wave is normally incident on a perfectly reflecting mirror. A glass photographic plate is placed on the mirror so that it makes a small angle α to the mirror. The photographic emulsion is almost transparent. But when it is later developed, a striped pattern is found due to the action of the wave. Predict the position and spacing of the black stripes which appear on the developed "negative" plate. Ignore any attenuation or reflection due to the glass photographic plate itself.

b) The experiment is repeated for incident waves making angle 45° to the normal to the mirror. Now what is the pattern of blackening on the negative? Distinguish the cases of light polarized with **E** parallel and perpendicular to the plane of incidence (i.e., the scattering plane).

Problem 2.9.   An infinitely long, thin-walled circular cylinder of radius b is split into two half cylinders. The upper half is fixed at potential $V = +V_0$ and the lower half at $V = -V_0$.

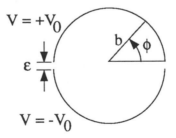

Figure 2.6.

a) Find the potential inside and outside the cylinder.

b) Calculate the charge density as a function of ϕ. (If your answer is in the form of an infinite sum, calculate this sum.)

c) Find the capacitance per unit length of the device when the two half cylinders are a distance ϵ apart at the rims ($\epsilon \ll b$).

Problem 2.10.   An electron is released from rest at a large distance r_0 from a nucleus of charge Ze and then "falls" toward the nucleus. For what follows, assume that the electron's velocity is such that $v \ll c$ and that the radiative reaction force on the electron is negligible.

a) What is the angular distribution of the emitted radiation?

b) How is the emitted radiation polarized?

c) What is the radiated power as a function of the separation between the electron and the nucleus?

d) What is the total energy radiated by the time the electron is a distance $r < r_0$ from the nucleus?

Chapter 3

Quantum Mechanics

Problem 3.1. A particle of mass m interacts in three dimensions with a spherically symmetric potential of the form

$$V(\mathbf{r}) = -c\delta(|\mathbf{r}| - a). \tag{3.1}$$

In other words, the potential is a delta function that vanishes unless the particle is precisely a distance a from the center of the potential. Here c is a positive constant.

a) Find the minimum value of c for which there is a bound state.

b) Consider a scattering experiment in which the particle is incident on the potential with a low velocity. In the limit of small incident velocity, what is the scattering cross-section? What is the angular distribution?

Problem 3.2. A particle of mass M bounces elastically between two infinite plane walls separated by a distance D. The particle is in its lowest possible energy state.

a) What is the energy of this state?

16

b) The separation between the walls is slowly (i.e., adiabatically) increased to $2D$.

i) How does the expectation value of the energy change?
ii) Compare this energy change with the result obtained classically from the mean force exerted on a wall by the bouncing ball.

c) Now assume that the separation between the walls is increased rapidly, with one wall moving at a speed $\gg \sqrt{E/M}$. Classically there is no change in the particle's energy since the wall is moving faster than the particle and cannot be struck by the particle while the wall is moving.

i) What happens to the expectation value of the energy quantum-mechanically?
ii) Compute the probability that the particle is left in its lowest possible energy state.

Problem 3.3. Consider a particle of charge e and mass m in constant, crossed **E** and **B** fields:

$$\mathbf{E} = (0, 0, E), \quad \mathbf{B} = (0, B, 0), \quad \mathbf{r} = (x, y, z). \qquad (3.2)$$

a) Write the Schrödinger equation (in a convenient gauge).

b) Separate variables and reduce it to a one-dimensional problem.

c) Calculate the expectation value of the velocity in the x-direction in any energy eigenstate (sometimes called the drift velocity).

Problem 3.4.   A particle of mass m and charge q sits in a harmonic oscillator potential $V = k(x^2 + y^2 + z^2)/2$. At time $t = -\infty$ the oscillator is in its ground state. It is then perturbed by a spatially uniform time-dependent electric field

$$\mathbf{E}(t) = Ae^{-(t/\tau)^2}\hat{\mathbf{z}} \qquad (3.3)$$

where A and τ are constants. Calculate in lowest-order perturbation theory the probability that the oscillator is in an excited state at $t = \infty$.

Problem 3.5.   Consider an elastic scattering experiment $a + X \to a + X$ with a and X having zero spin, and X much heavier than a. The total cross-section σ_{tot} as a function of momentum $\hbar k$ behaves as shown in Figure 3.1. A contribution to σ_{tot} from a resonance is observable at

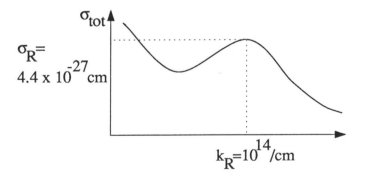

Figure 3.1.

all angles except 90°, where the contribution vanishes. Far off resonance σ_{tot} is isotropic.

a) What is the angular momentum J of the resonance?

b) Calculate the approximate value of the differential cross-section at resonance at a scattering angle of 180°.

Problem 3.6. a) A spin 1/2 electron is in a uniform magnetic field $\mathbf{B_0} = B_0\hat{z}$. At time $t = 0$ the spin is pointing in the x-direction, i.e., $\langle S_x(t = 0)\rangle = \hbar/2$. Calculate the expectation value $\langle \mathbf{S}(t)\rangle$ at time t.

b) An additional magnetic field $\mathbf{B_1} = \frac{1}{2}B_1[\cos(\omega t)\hat{x} + \sin(\omega t)\hat{y}]$ is now applied. If an electron in the combined field $\mathbf{B_0} + \mathbf{B_1}$ has spin pointing along $+\hat{z}$ at time $t = 0$, what is the probability that it will have flipped to $-\hat{z}$ at time t?

Problem 3.7. Pion-nucleon scattering at low energies can be qualitatively described by an effective interaction potential of the form:

$$V = \left(\frac{a^2}{4\pi}\right)\left(\frac{e^{-\mu r}}{r}\right)\mathbf{I}^{(\pi)}\cdot\mathbf{I}^{(N)}. \tag{3.4}$$

Here a and μ are constants, r is the relative pion-nucleon coordinate, and $\mathbf{I}^{(\pi)}$ and $\mathbf{I}^{(N)}$ are the pion and nucleon isospin operators.

a) Calculate the ratio of the scattering cross-sections with total isospin $I = 3/2$ and $I = 1/2$.

b) Calculate in the Born approximation the low-energy total cross-sections for the reactions:

$$\pi^+ + p \rightarrow \pi^+ + p,$$
$$\pi^- + p \rightarrow \pi^- + p,$$
$$\pi^- + p \rightarrow \pi^0 + n.$$

NB: If you are not familiar with isospin, you may consider the two particles to have (ordinary) spin 1 and spin 1/2 with a spin-spin interaction and initial and final states which are eigenstates of S_z for each particle. The corresponding S_z values are:

$$\begin{pmatrix} \pi^+ \\ \pi^0 \\ \pi^- \end{pmatrix} = \begin{pmatrix} 1 \\ 0 \\ -1 \end{pmatrix} \quad \text{and} \quad \begin{pmatrix} p \\ n \end{pmatrix} = \begin{pmatrix} 1/2 \\ 1/2 \end{pmatrix}. \tag{3.5}$$

Problem 3.8. A particle of total energy $E = \hbar^2\alpha^2/2m$ moves in a series of N contiguous one-dimensional regions. The potential in the nth region is

$$V_n = -(n^2 - 1)E, \quad \text{where n} = 1, 2, \ldots, N. \tag{3.6}$$

All regions are of equal width π/α except for the first and the last, which are of effectively infinite extent. Calculate the two transmission coefficients for a particle incident from either end.

Problem 3.9.   A neutron beam is polarized parallel to a uniform magnetic field **B**. The beam is then split into two halves. One beam continues through the uniform field; the other beam passes through a field with the same fixed magnitude but which changes *gradually* in direction. The two beams are recombined and the intensity is measured. The path lengths are equal so that the interference would be constructive if **B** were uniform. Assume the neutrons in one beam experience, in their rest frame, a time-dependent magnetic field (sketched in Figure 3.2), given by

$$\mathbf{B} = B_0 \left[\sin\theta \cos\phi(t)\hat{\mathbf{x}} + \sin\theta \sin\phi(t)\hat{\mathbf{y}} + \cos\theta\hat{\mathbf{z}} \right], \tag{3.7}$$

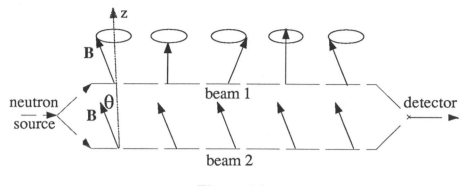

Figure 3.2.

where the angle $\phi(t)$ varies adiabatically from zero to 2π. Those in the other beam experience a constant field:

$$\mathbf{B} = B_0 \left[\sin\theta \hat{\mathbf{x}} + \cos\theta \hat{\mathbf{z}} \right]. \tag{3.8}$$

a) Calculate the ground-state wavefunction, *in a basis of S_z eigenstates*, for a constant **B**-field pointing in an arbitrary direction labeled as above by θ and ϕ.

b) Calculate, in the adiabatic approximation, the relative phase change of the two neutron beams after they have passed through the magnetic fields, and hence the intensity of the recombined beam as a function of θ.

Problem 3.10. An isolated hydrogen atom has a hyperfine interaction between the proton and electron spins ($\mathbf{S}_1$ and $\mathbf{S}_2$, respectively) of the form $J\mathbf{S}_1 \cdot \mathbf{S}_2$. The two spins have magnetic moments $\alpha\mathbf{S}_1$ and $\beta\mathbf{S}_2$, and the system is in a uniform static magnetic field **B**. Consider only the orbital ground state.

a) Find the exact energy eigenvalues of this system and sketch the hyperfine splitting spectrum as a function of magnetic field.

b) Calculate the eigenstates associated with each level.

Chapter 4

Thermodynamics & Statistical Mechanics

Problem 4.1.   Consider a system of $N \gg 1$ non-interacting particles in which the energy of each particle can assume two and only two distinct values: 0 and E ($E > 0$). Denote by n_0 and n_1 the occupation numbers of the energy levels 0 and E, respectively. The fixed total energy of the system is U.

a) Find the entropy of the system.

b) Find the temperature as a function of U. For what range of values of n_0 is $T < 0$?

c) In which direction does heat flow when a system of negative temperature is brought into thermal contact with a system of positive temperature? Why?

———

Problem 4.2.   Consider a heteronuclear diatomic molecule with moment of inertia I. In this problem, only the rotational motion of the molecule should be considered.

a) Using classical statistical mechanics, calculate the specific heat $C(T)$ of this system at temperature T.

b) In quantum mechanics, this system has energy levels

$$E_j = \frac{\hbar^2}{2I} j(j+1) \quad j = 0, 1, 2, \ldots \tag{4.1}$$

Each j level is $(2j + 1)$-fold degenerate. Using quantum statistical mechanics, find expressions for the partition function Z and the average energy $\langle E \rangle$ of this system, as a function of temperature. Do not attempt to evaluate these expressions.

c) By simplifying your expressions in (b), derive an expression for the specific heat $C(T)$ that is valid at very low temperatures. In what range of temperatures is your expression valid?

d) By simplifying your answer to (b), derive a high-temperature approximation to the specific heat $C(T)$. What is the range of validity of your approximation?

Problem 4.3. A simplified model of diffusion consists of a one-dimensional lattice, with lattice spacing a, in which an "impurity" makes a random walk from one lattice site to an adjacent one, making jumps at time intervals τ.

a) After N jumps have been made, find the probability that the atom has moved a distance d from its starting point, in the limit of large N.

b) The diffusion coefficient is defined by the differential equation

$$D\frac{\partial^2 f}{\partial x^2} = \frac{\partial f}{\partial t}, \tag{4.2}$$

where f is the concentration of the impurity. Find an expression for D in the model described above.

Problem 4.4.   The temperature of a long vertical column of a particular substance is T everywhere. Below a certain height $h(T)$ the substance is solid, whereas above $h(T)$ it is in a liquid phase. Calculate the density difference $\Delta\rho = \rho_s - \rho_l$ between the solid and liquid ($|\Delta\rho| \ll \rho_s$), in terms of L (the latent heat of fusion per unit mass), dh/dT, T, and g, the acceleration due to gravity.

Problem 4.5.   The operation of a gasoline engine is (roughly) similar to the Otto cycle (Figure 4.1):

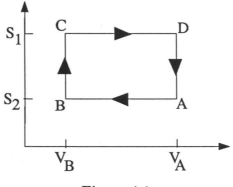

Figure 4.1.

$A \to B$ Gas compressed adiabatically
$B \to C$ Gas heated isochorically (constant volume; corresponds to combustion of gasoline)
$C \to D$ Gas expanded adiabatically (power stroke)
$D \to A$ Gas cooled isochorically.

Compute the efficiency of the Otto cycle for an ideal gas (with temperature-independent heat capacities) as a function of the compression ratio V_A/V_B, and the heat capacity per particle C_V.

Problem 4.6. Calculate the drag force on a disk of radius R moving with constant velocity $\mathbf{V}$ (perpendicular to the plane of the disk) in a rarefied gas of density n that is in thermal equilibrium at temperature T. Assume that the gas molecules collide elastically with the disk, that the speed of the disk is slow compared with the average molecular speed, and that the disk is large compared to a molecule but small compared to the mean free path of the molecules.

Problem 4.7. A wire of length l and mass per unit length μ is fixed at both ends and tightened to a tension τ. What is the rms fluctuation, in classical statistics, of the midpoint of the wire when it is in equilibrium with a heat bath at temperature T? A useful series is

$$\sum_{m=0}^{\infty} (2m+1)^{-2} = \frac{\pi^2}{8}. \tag{4.3}$$

Problem 4.8. Consider a vapor (dilute gas) in equilibrium with a submonolayer (i.e., less than one atomic layer) adsorbed film. Model the binding of atoms to the surface with a potential energy $(-\epsilon_0)$. Assume there are N possible sites for adsorption and find the vapor pressure as a function of surface concentration, n/N.

Problem 4.9. Consider a type I superconducting material with a parabolic coexistence curve separating the (uniform) superconducting

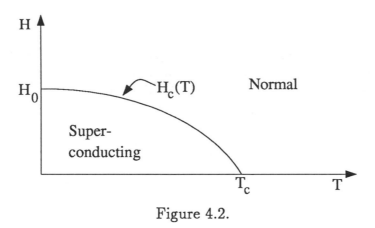

Figure 4.2.

and normal phases (see Figure 4.2). H is the external magnetic field, and T is the temperature. Ignore the tiny magnetization of the normal phase. The critical field H_c is given by

$$H_c = H_0 + aT + bT^2. \tag{4.4}$$

a) Why must the coefficient a be zero?

b) Calculate the latent heat per unit volume as a function of T along the coexistence curve in terms of H_0 and T_c, as shown in Figure 4.2.

c) Calculate the discontinuity in the specific heat per unit volume at constant H along the coexistence curve.

Problem 4.10.   A monatomic crystal consists of $N \approx 10^{23}$ atoms which may be situated in either of two types of potential wells: "normal" positions, labeled by "O," and "interstitial" positions, labeled by "X" (Figure 4.3). The O- and X-sites are arranged on interpenetrating cubic lattices, and the energy of an atom in position X is greater than that of an atom in position O by an amount ϵ.

a) Calculate the entropy for states with n atoms in the interstitial sites when $1 \ll n \ll N$, assuming any atom originally on an O-site will hop only to one of the eight nearest interstitial sites.

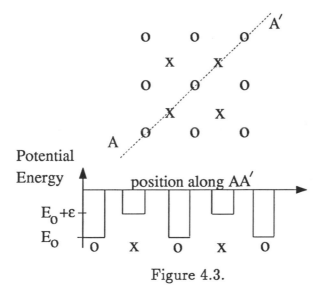

Figure 4.3.

b) Repeat the calculation of the entropy in (a), assuming that the n unoccupied sites ("holes") of the O-lattice are uncorrelated with the n occupied interstitial sites.

c) Calculate the fraction of intersititial sites that are occupied at low temperatures $k_B T \ll \epsilon$ for (a) and (b) above. Neglect interactions between atoms on different sites.

Chapter 5

Condensed Matter Physics

Problem 5.1.   The tunneling of quasiparticles across a thin oxide barrier separating two metals (Figure 5.1) can be described by a tunneling amplitude M which is independent of the initial and final state energies.

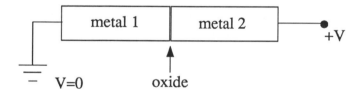

Figure 5.1.

a) Derive an expression for the conductance of the junction in the low temperature $(T \ll E_F)$ and low voltage $(eV \ll E_F)$ limits when both metals are normal (i.e., not superconducting).

b) Now consider the case in which metal 1 is a superconductor with energy gap Δ and metal 2 is normal. Calculate the $T = 0$ current as a function of voltage.

c) In a Pb-oxide-Al junction, both metals are superconducting with $\Delta_{Pb} > \Delta_{Al}$. The I-V characteristic shows a maximum and minimum at $V_1 = 11.8 \times 10^{-4}$ volts and $V_2 = 15.2 \times 10^{-4}$ volts (see Figure 5.2). Derive the values of the energy gaps Δ_{Pb} and Δ_{Al} from these data.

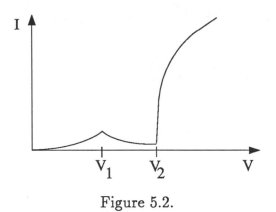

Figure 5.2.

Problem 5.2. Type II superconductors undergo a second-order transition from the normal to superconducting phases in a uniform magnetic field. The phase diagram is sketched in Figure 5.3. Calculate the

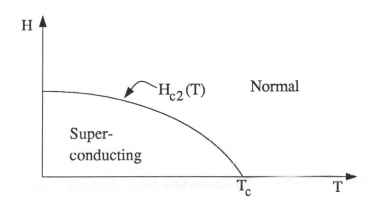

Figure 5.3.

phase boundary line $H_{c2}(T)$ assuming the Landau-Ginzburg equation

for the superconducting order parameter $\psi(\mathbf{x})$ is valid at all temperatures $T < T_c$ (i.e., the zero-field transition temperature). Recall that

$$\left(\nabla - i\frac{2e}{\hbar c}\mathbf{A}\right)^2 \psi + \xi^{-2}\psi + \beta|\psi|^2\psi = 0, \qquad (5.1)$$

where $\xi(T) = \xi_0(1 - T/T_c)^{-1/2}$ is the coherence length, β is a constant,
and $\mathbf{A}(\mathbf{x})$ is the vector potential.

Problem 5.3. a) A model for the low-temperature properties of glass
proposes that in the random network of bonds there are many places
where atoms have two alternating local, metastable equilibrium configurations, separated by an energy difference E_α (Figure 5.4). These
"two-level" centers are assumed to be noninteracting and it is also as-

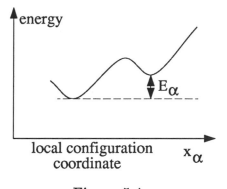

Figure 5.4.

sumed that the low-frequency linear vibrations around either configuration are described by ordinary phonons. Find the leading contribution
to the low-temperature specific heat. Assume E_α is a random variable
with a flat probability distribution $P(E_\alpha)$.

b) In what other system is the temperature dependence of the specific heat similar to that found in (a)? Estimate the relative magnitudes
of the two results. Assume for part (a) that the density of levels for the
"two-level" centers is 1 state/(eV atom).

c) In one form of crystalline carbon the low-temperature lattice specific heat is found to be proportional to T^2. What does this say about the structure of this form of carbon? Explain your answer.

Problem 5.4. Barium titanate is an ionic crystal that exhibits a ferroelectric transition at $T_c = 381$ K. Above the Curie temperature, $BaTiO_3$ has cubic symmetry with a unit cell shown in Figure 5.5a. Below T_c, $BaTiO_3$ develops a spontaneous electric polarization **P**.

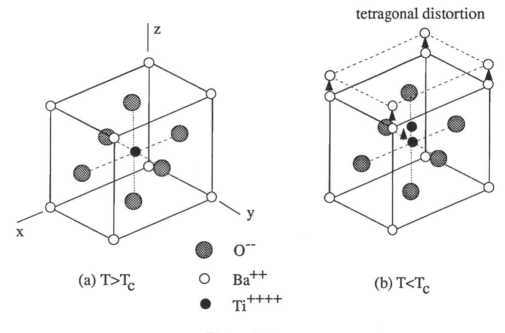

tetragonal distortion

(a) T>T_c ⬤ O^{--}

           ◯ Ba^{++} (b) T<T_c

           ● Ti^{++++}

Figure 5.5.

a) Construct the Landau free energy function for the ferroelectric transition through fourth order in the order parameter **P**.

b) What are the allowed directions of **P**, relative to the crystal axes, if the transition is second order?

c) Sketch a plausible form for the phase diagram for $BaTiO_3$ in the pressure-temperature plane based on the free energy function constructed in (a). Label the order of all transition lines.

d) The appearance of the electric polarization $\mathbf{P}$ for $T < T_c$ must be accompanied by displacements of the symmetric arrangement of the ions. Thus, ferroelectricity is not simply described by a single order parameter; the free energy function must also depend upon the *strain*. Assuming that the distorted lattice has tetragonal symmetry (Figure 5.5b), show that the coupling of the electric polarization and the strain leads to a first-order transition if the crystal is sufficiently "soft" (i.e., the elastic constant is less than some critical value).

Problem 5.5.   Give short answers for the following.

a) Explain how to construct a refrigerator from strips of Cu and Sn.

b) The heat capacity of crystalline EuO varies as $C_V = aT^{3/2} + bT^3$ for $0 \le T \le 70$ K. What state of matter is EuO most likely to be?

c) The B (magnetic induction) vs. H (magnetic field) curve for superconducting V_3Ga is shown in Figure 5.6. What is the distance between flux lines for $H = 2.3 \times 10^3$ gauss?

d) A long pipe filled with a powder from a simple cubic crystal of lattice constant b is placed in the path of a neutron beam from a reactor. The powder acts as a filter. What are the wavelengths λ of neutrons which are transmitted? Why?

e) A magnetic field, $H = 100$ kgauss, is perpendicular to a strip of Cu as shown in Figure 5.7. A uniform current of $I = 0.1$ A flows steadily through the strip, and the voltage across points A and B is measured to be $V = 8.1 \times 10^{-8}$ volts. Use this data to estimate the density of charge carriers n in Cu.

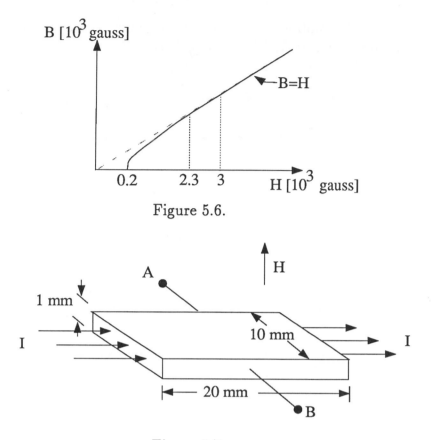

Figure 5.6.

Figure 5.7.

f) Sketch the energy-loss spectrum for electrons reflected from a Mg film. The incident electron energy is 2 keV, and the electron density is $n = 8.6 \times 10^{22}$ cm^{-3}.

Problem 5.6. The energy–momentum relation for elementary excitations in superfluid ^{4}He has the form shown in Figure 5.8. The long wavelength excitations are phonons with $\hbar\omega_q = cq$, where c is the speed of sound and $q = |\mathbf{q}|$ is the magnitude of the wavevector. For shorter wavelengths the excitation spectrum has a local minimum; the excitations near the minimum are known as rotons and their dispersion may be approximated by $\hbar\omega_q \approx \Delta + \alpha(q - q_0)^2$.

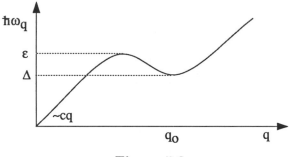

Figure 5.8.

a) Sketch the density of states $\rho(E)$ for single excitations throughout the energy range indicated in the dispersion relation above. Indicate the nature of any singularities.

b) Light-scattering experiments have revealed that the lowest energy state of a pair of rotons is less than 2Δ. Assume that rotons interact with each other through an attractive contact interaction $V(\mathbf{r}_1 - \mathbf{r}_2) = -g\delta^{(3)}(\mathbf{r}_1 - \mathbf{r}_2)$, and calculate the dependence of the binding energy on the coupling strength g for a pair of rotons in the limit of small positive g. Is there a critical value of g for the appearance of a bound state? Hint: Consider the Schrödinger equation in momentum space for the wavefunction of two rotons; a cutoff may be needed.

Problem 5.7.   Molecular liquids often undergo a phase transition from an isotropic liquid state to a liquid state which exhibits uniaxial anisotropy in its dielectric, magnetic and flow properties (called the nematic phase). The susceptibility tensor $\bar{\chi}$ has the form $\chi_{ij} = \chi_0\delta_{ij} + \chi_a n_i n_j$, where the unit vector $\hat{n}$ defines the axis along which the molecules are aligned (on average), and $\chi_a > 0$. Although the molecules exhibit no positional ordering in the nematic phase, there is an elastic energy associated with spatial curvature of the anisotropy axis $\hat{n}$ given by

$$F_{elastic} = \frac{k}{2} \int dV \sum_{i,j} \left(\frac{\partial n_j}{\partial x_i} \right)^2, \tag{5.2}$$

with elastic constant $k > 0$.

Consider a nematic liquid confined between two glass slides which are prepared (brushed) so that molecules at the surface of the glass orient themselves parallel to the surface along the x-direction (see Figure 5.9). A magnetic field $\mathbf{H} = H\hat{y}$ (where $\hat{y}$ is directed out of the page) is applied, and it is observed by light scattering that a transition from the uniform nematic phase, with $\hat{n} = \hat{x}$ everywhere, to a spatially varying phase, with

$$\hat{n}(z) = \cos \theta(z)\hat{x} + \sin \theta(z)\hat{y}, \tag{5.3}$$

occurs for $H > H_c$, where $\hat{z}$ is perpendicular to the slides.

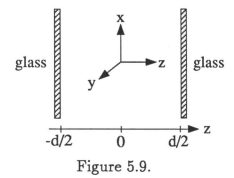

Figure 5.9.

a) What characteristic length determines the scale for spatially varying states described by $\theta(z)$?

b) Determine the critical field H_c at which the distortion of $\hat{n}$ first appears from the differential equation for $\theta(z)$.

c) How does the maximum distortion of $\hat{n}$ vary with magnetic field H for $H \gg H_c$?

Problem 5.8. An electron of mass m in a classical deformable elastic medium with bulk modulus B forms a self-trapped, or polaronic, state if the elastic energy of a local lattice distortion is less than the quantum-mechanical binding energy of the electron to such a distortion. For a lattice distortion with a wavelength large compared to the interatomic spacing a, the electron feels a potential energy of the form $V(\mathbf{x}) = \epsilon_0 \nabla \cdot \mathbf{U}(\mathbf{x})$, where $\mathbf{U}(\mathbf{x})$ is the displacement field of the atoms from equilibrium, and ϵ_0 is some constant.

a) Consider a fluctuation of linear size L in d spatial dimensions. For simplicity assume the potential felt by the electron is a square well. Determine, to within a numerical constant of order unity, the lattice distortion $\nabla \cdot \mathbf{U}(\mathbf{x})$ that minimizes the total energy of an electron in the elastic field.

b) Compute the total energy $E(L)$ as a function of L, the size of the polaron. Sketch $E(L)$ for $d = 1$ and $d = 3$ dimensions; pay careful attention to the asymptotic behaviors for large and small L.

c) The smallest dimension of a polaron is fixed by the spacing a between ions in the lattice. Assuming that the elastic energy becomes infinite for square-well distortions with $L \leq a$, determine the smallest value of ϵ_0 for which the polaron is the true ground state of the system for $d = 3$ dimensions, in terms of m, B, a, and any fundamental constants needed.

d) How does the answer to part (c) change when $d = 1$?

––––––––––––––––––––––––––

Problem 5.9. Consider the interacting electron hamiltonian

$$H = \sum_{k,s} \epsilon_{ks} c_{ks}^\dagger c_{ks} + \frac{1}{2} \sum_{k,k',q,s,s'} V(q) c_{(k+q)s}^\dagger c_{(k'-q)s'}^\dagger c_{k's'} c_{ks'} \qquad (5.4)$$

of a one-dimensional free electron gas in a "volume" L with one-electron energies

$$\epsilon_{ks} = \frac{\hbar^2 k^2}{2m}. \qquad (5.5)$$

Here $s = \pm 1/2$ is the spin index.

a) For the noninteracting case $V(q) = 0$, evaluate the total kinetic energy at zero temperature for:
i) a ferromagnetic state with all spins up, and
ii) a paramagnetic state with half of the spins up and half down.
Express your results in terms of the number density $n = N/L$ where N is the total number of electrons.

In the Hartree-Fock approximation the interaction term in the hamiltonian (5.4) is replaced by

$$V_{HF} = \frac{1}{2} \sum_{k,k',s,s'} V(0) n_{k's'} c_{ks}^\dagger c_{ks} - \frac{1}{2} \sum_{k,k',s} V(k - k') n_{k's} c_{ks}^\dagger c_{ks} . \qquad (5.6)$$

Note that the second term above is not summed over s'. Here $n_{ks} = < c_{ks}^\dagger c_{ks} >$ is the occupation of the orbital ks.

For the following, consider a short range repulsive potential $V(r) = G\delta(r)$ with the Fourier transform $V(q) = G$.

b) What are the single particle energies ϵ'_{ks} in this approximation?

c) Evaluate the ground-state energy E^{HF} for the ferromagnetic and paramagnetic states in this approximation. Plot

$$\frac{\Delta E}{N} = \frac{E^{HF}(\text{ferro}) - E^{HF}(\text{para})}{N} \qquad (5.7)$$

as a function of n. Find the critical density n_c at which a transition to a ferromagnetic state will occur.

Problem 5.10. Electrons in the surface "inversion" layer of a crystal of silicon covered with an oxide layer behave like a purely two-dimensional electron gas with effective mass m^*. Ignore electron-

electron interactions and assume that the electron spin g-factor is small: $g_s \ll m/m^*$.

a) Write down the eigenenergies in a perpendicular magnetic field B, and the lowest energy eigenfunctions. What are the degeneracies of the eigenvalues in a large system of area $\mathcal{A}$? (Hint: work in the gauge where the vector potential is $A_y = Bx, A_x = 0$. Consider the position of the eigenfunctions.)

b) A counter-electrode on the other side of the oxide is positively charged with an external voltage $V = Q/C$ (with C the capacitance). Sketch the Fermi level ϵ_F in the inversion layer as a function of V at zero temperature with appropriate scales shown on the axes. Sketch how this result would be affected by a small amount of disorder, for low temperature.

c) Sketch the Hall conductance, $\sigma_H = j/E_H$, (with j the two-dimensional current density and E_H the Hall electric field) as a function of ϵ_F, at low temperature, with appropriate scales on the axes.

Chapter 6

Relativity & Astrophysics

Problem 6.1.   For each of the following, state the first non-trivial approximation to the relativistic shift in the observed frequency or time scale relative to the proper frequency or time:

a) the halflife of a muon created in flight with momentum $p \gg mc$,

b) the Mossbauer resonance frequency of nuclei of mass m as a function of temperature T,

c) Mossbauer resonant frequencies compared at the source and at rest at altitude H above the source in the gravitational field g,

d) the age of a cosmonaut who spends several years in a low earth orbit, and

e) the time between explosions in a galaxy observed at a redshift $z \ll 1$.

Problem 6.2.   A rocket ship accelerates away from the earth at constant acceleration g (in the rocket's frame).

a) Calculate the angular size of the earth as viewed from the rocket as a function of proper time on the rocket.

b) The angular size in part (a) approaches a limit as proper time gets large. What is it?

Problem 6.3. An experimentalist has an idea to test gravitational magnetism. A very long cylinder of uniform density ρ and radius R is spun clockwise rapidly on its axis with angular velocity ω. Light travels around the light pipe of radius r. Find an approximate formula, valid to lowest order in ω, for the time difference $\Delta t = t_{cw} - t_{ccw}$ where t_{cw} (t_{ccw}) is the proper time for an observer located at O for light to travel clockwise (counterclockwise) around the cylinder. Assume $r \approx R$ and $G\rho R^2 \ll 1$. Give a numerical estimate of the ratio between this time

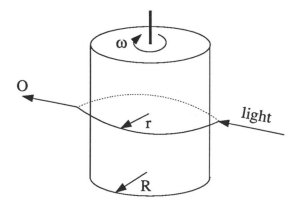

Figure 6.1.

difference and the period of optical light, for an apparatus which could fit in a room.

Problem 6.4. The metric for an isotropic universe in comoving coordinates is given by the Robertson-Walker form,

$$ds^2 = -dt^2 + R^2(t)\left[\tilde{g}_{rr}dr^2 + \tilde{g}_{\theta\theta}d\theta^2 + \tilde{g}_{\phi\phi}d\phi^2\right], \qquad (6.1)$$

where $\tilde{g}_{rr} = (1 - kr^2)^{-1}$, $\tilde{g}_{\theta\theta} = r^2$, and $\tilde{g}_{\phi\phi} = r^2 \sin^2\theta$ are the non-zero components of the metric for a three-dimensional maximally symmetric space. In this problem you may take $k = 1$ corresponding to a closed universe. With this metric the non-zero components of the Ricci tensor are

$$- R_{tt} = 3\ddot{R}/R \quad \text{and} \quad R_{ij} = (R\ddot{R} + 2\dot{R}^2 + 2k)\tilde{g}_{ij}, \qquad (6.2)$$

where dots denote time derivatives.

As a simple model of the early universe assume that the universe contains only a thermal gas of photons at temperature T, with energy density $\rho = aT^4$, and entropy density $S = (4/3)aT^3$, where $a = \pi^2 k_B^4/15c^3\hbar^3$ is the black-body constant. Define the current age of the universe to be t_0.

a) Determine $T_0 \equiv T(t_0)$ and $R_0 \equiv R(t_0)$ in terms of the Hubble constant $H_0 \equiv (\dot{R}/R)_{t_0}$ and the deceleration parameter $q_0 \equiv -(R\ddot{R}/\dot{R}^2)_{t_0}$.

b) Determine t_0 in terms of q_0 and H_0 and find $R(t)$ and $T(t)$ for very early times.

c) Determine the radius L_1 of a region at time $t_1 \ll t_0$ which will evolve into a region of radius $L_0 \equiv t_0$ at time t_0. How many causally-connected regions were contained in a volume L_1^3 at time t_1? Give a numerical estimate assuming $H_0 = 100\,\text{km s}^{-1}\,\text{Mpc}^{-1}$, $q_0=2$, and $k_B T(t_1)$ is 10^{15} GeV. (Note: 1 pc=3.26 light years.)

d) In order to reconcile the answer to (c) with the observed homogeneity of the universe, A. Guth has proposed that the universe underwent an "inflationary" phase at very early times resulting in a departure from adiabaticity. By what factor must the resulting entropy have increased in order to solve this "homogeneity problem"?

Problem 6.5. Consider the Schwarzschild metric for a black hole of mass M:

$$ds^2 = -\left(1 - \frac{2M}{R}\right) dt^2 + \left(1 - \frac{2M}{R}\right)^{-1} dr^2 + r^2(d\theta^2 + \sin^2\theta d\phi^2) . \quad (6.3)$$

A particle of mass m is in a stable circular orbit of radius $R > 2M$. The momentum four-vector is p^μ.

a) Show that p_t and p_ϕ are constants of the motion. Then establish an "energy" conservation equation for radial motion in the equatorial plane, and derive the relationship between p_ϕ, M and R.

b) Compute the orbital period as indicated on a clock carried along with the particle.

c) At the completion of each orbit the particle emits a photon. This is eventually received by an observer O located very far away. What is the period as perceived by O?

d) An observer O' is fixed at radius R on the orbit, held there by a rocket motor. What is the period of the particle as measured by O'?

Problem 6.6. Consider a rocket which is using its motor to hold itself *at rest* at radius R (in Schwarzschild coordinates) away from a point star of mass M.

a) Although it is at coordinate rest, it is accelerating with respect to any local inertial frame. What is the invariant acceleration?

b) To maintain this acceleration the rocket has to burn fuel. Suppose it gets this thrust by producing perfectly collimated downward-moving photons. How does the rocket's total mass depend on proper time?

c) Answer part (a) for a rocket which is holding itself in a circular orbit with proper angular frequency ω: i.e., what is the magnitude of the invariant acceleration? Is there a choice of ω for which the invariant acceleration goes to zero?

Problem 6.7. A domain wall is a planar generalization of a cosmic string: it has stresses and energy densities concentrated in a plane and an energy-momentum tensor independent of boosts or translations in that plane. The following metric is a solution to Einstein's equations for a domain wall lying in the $z = 0$ plane:

$$ds^2 = -(1 - k|z|)^2 dt^2 + dz^2 + (1 - k|z|)^2 e^{2kt}(dx^2 + dy^2) \qquad (6.4)$$

(the constant k is proportional to the energy per unit area).

a) Away from the wall the energy-momentum tensor, and therefore $R_{\mu\nu}$, must vanish. Show that the component R^x_{zxz} of the Riemann tensor vanishes for $z \neq 0$. Actually *all* components vanish for $z \neq 0$, but we won't ask you to show it.

b) Despite the fact that space is flat, particle geodesics have some peculiar features. Show that a particle placed at rest a small distance to one side of the domain wall accelerates *away* from the wall and calculate the initial magnitude of the invariant acceleration.

Problem 6.8. The spin of a gyroscope is characterized by a classical spin four-vector satisfying $S \cdot U = 0$ (where $U^\mu = dx^\mu/d\tau$ is the center-of-mass four-velocity of the gyroscope) and $S \cdot S = $ constant. If the gyroscope moves along a geodesic orbit, the spin vector evolves by parallel transport. Consider the case of a gyroscope moving in a stable circular geodesic orbit of radius R in a Schwarzschild metric,

$$ds^2 = -\left(1 - \frac{2M}{r}\right) dt^2 + \frac{dr^2}{1 - 2M/r} + r^2 d\Omega^2. \qquad (6.5)$$

a) Convert the parallel transport equation into an equation for the evolution, as a function of azimuthal angle around the orbit, of the spatial components of the spin.

b) Solve this equation to find the rate of precession of the spin vector as seen by an observer fixed at a given point on the orbit.

Problem 6.9. Consider a dust particle in orbit around the sun. Ignore the effects of the solar wind and magnetic field.

a) Estimate the minimum size that a dust particle may have to avoid being blown out of the solar system.

b) Particles larger than the minimum size you estimated above will spiral into the sun. What is the source of the drag? Estimate the lifetime of such a particle in orbit at the distance of earth.

Useful numbers:

$$
\begin{aligned}
M_\odot &= 2 \times 10^{33} \text{ g} \\
R_{\text{earth}} &= 1.5 \times 10^{13} \text{ cm} \\
G &= 6.67 \times 10^{-8} \text{ergs cm/g}^2 \\
L_\odot &= 4 \times 10^{33} \text{ ergs/s.}
\end{aligned}
$$

Problem 6.10. A model for the universe, first proposed by Kasner, has the metric

$$ds^2 = -dt^2 + t^{2\alpha}dx^2 + t^{2\beta}dy^2 + t^{2\gamma}dz^2,$$

where α, β, and γ are constants.

a) Use the vacuum Einstein equations to find relations between α, β, and γ.

b) Find all the Killing vectors of this spacetime for general α, β, and γ.

c) Comment briefly on why you feel this cannot be made into a realistic cosmological model.

d) In the case $\alpha = 1$, $\beta = \gamma = 0$, show that the spacetime is flat.

[You might find it useful to recall that the Riemann tensor is

$$R^\alpha{}_{\beta\gamma\epsilon} = \Gamma^\alpha{}_{\epsilon\beta,\gamma} - \Gamma^\alpha{}_{\gamma\beta,\epsilon} + \Gamma^\alpha{}_{\rho\gamma}\Gamma^\rho{}_{\epsilon\beta} - \Gamma^\alpha{}_{\rho\epsilon}\Gamma^\rho{}_{\beta\gamma},$$

where $\Gamma^\alpha{}_{\beta\gamma}$ is the Christoffel symbol.]

Chapter 7

Nuclear Physics

Problem 7.1. A proposed geologic solar neutrino detector utilizes ^{205}Tl as the target material (70% of natural thallium). Thallium-bearing ore would be searched for ^{205}Pb ($t_{1/2} = 14$ million years), created by neutrino captures.

a) The state structure in $A = 205$ is given in Figure 7.1. Explain the states in terms of a simple shell model.

b) In order to determine the cross-section for neutrino capture to the $1/2^-$ state of ^{205}Pb, one needs to look at the $A = 206$ system also shown in Figure 7.1. Again, explain the observed states.

c) Assume that the ^{205}Tl ground state is known to have a 10% admixture of the neutron configuration $(f_{5/2})^{-2} (p_{1/2})^2$. Then using the information provided, calculate the equilibrium number of ^{205}Pb atoms due to solar neutrino capture to the $1/2^-$ state, in ore containing a gram of ^{205}Tl. Approximate the neutrino spectrum by a single energy $E_\nu = \overline{E_\nu} = 0.26$ MeV.

Shell model: p: $1h_{11/2}$ $2d_{3/2}$ $3s_{1/2}$ (82) $1h_{9/2}$ $1i_{13/2}$
               n: $3p_{3/2}$ $2f_{5/2}$ $3p_{1/2}$ (126) $2g_{9/2}$ $1j_{15/2}$.

Solar constant $= 0.14$ W/cm^2.
$4p \rightarrow {}^4$He $+ 2e^+ + 2\nu + 26.74$ MeV.

46

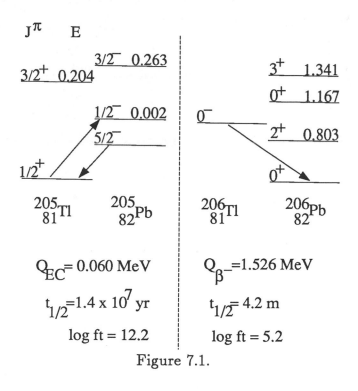

Figure 7.1.

Problem 7.2. a) Assuming the magnetic moment of an odd-A nucleus $\mu_I = g_I \mu_N I$ is determined by the odd nucleon, derive an expression for g_I for $I = j = l \pm 1/2$ (Schmidt limits).

b) What is the expected sequence of shell model energy levels up to $^{56}_{28}Ni_{28}$?

c) What does the Schmidt limit predict for the magnetic moment of $^{17}_{8}O_9$, $^{23}_{11}Na_{12}$, and $^{45}_{21}Sc_{24}$? Note: $\mu_p = 2.793\mu_N$ and $\mu_n = -1.913\mu_N$.

d) Calculate the electric quadrupole moments Q in the extreme single particle model for $^{15}_{7}N_8$ and $^{11}_{5}B_6$. Express your answer in terms of $\langle r^2 \rangle$ and then make an estimate of Q using reasonable nuclear sizes.

Problem 7.3.   The classic example of a superallowed pure Fermi transition is the beta decay of ^{14}O to the 2.3 MeV level of ^{14}N, as illustrated in Figure 7.2.

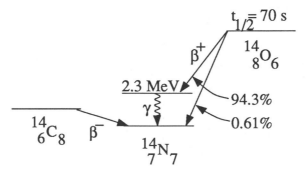

Figure 7.2.

a) What are the spin-parity J^π and isospin assignments for the initial and final states of this superallowed decay? Justify your answer.

b) What is the value of the Fermi matrix element for this decay?

c) Specify the spin-parity and isospin for the ground state of ^{14}N and classify the beta decays from the ground states of ^{14}C and ^{14}O to this level. State how the matrix elements for the β^- and β^+ compare.

d) There are several other nuclear mass triads for which superallowed pure Fermi decays occur (in particular for $A = 4n + 2$, $n = 2, 3, 4, \cdots$). Estimate the A dependence on the halflife for the superallowed decays by first estimating the dependence of the halflife on the beta end-point energy (neglect Coulomb effects on the outgoing β particle) and then the dependence of the end-point energy on the atomic number Z of the daughter nucleus. Note that in these cases $Z \approx A/2$.

e) The experimental halflife for the superallowed pure Fermi decay $^{54}_{27}\text{Co} \rightarrow ^{54}_{26}\text{Fe} + \beta^+ + \nu$ is 0.19 s. Make an estimate on the basis of part (d) and the halflife of ^{14}O and compare.

Problem 7.4. Neutrino Sources:

Atomic mass excesses (in MeV):

^1_0n	8.071
^1_1H	7.289
^2_1H	13.136
^4_2He	2.425
$^{208}_{82}\text{Pb}$	21.759
$^{232}_{90}\text{Th}$	35.447

Astronomy:
   solar luminosity at the earth's orbit $= 1.4 \text{ kW/m}^2$

Geology:
   thermal gradient of the earth $= 30 \text{ K/km}$
   thermal conductivity of granite $= 2.8 \text{ kcal/m hr K}$

A. Assume the sun derives its energy from converting hydrogen to helium.

a) What is the neutrino flux at the earth's surface due to the sun? (A numerical result in particles/m²s is required.)

b) Derive the shape of the neutrino spectrum, i.e., the flux versus neutrino energy.

c) What percentage of the sun's energy production is dissipated by neutrino emission?

B. The earth also produces energy from radioactive decay of heavy elements and thus has a molten core.

a) What type of neutrinos are produced?

b) Estimate the neutrino flux at the surface of the earth.

C. Consider a large fission power reactor.

a) What type of neutrinos are emitted?

b) Estimate the neutrino flux at 100 meters distance from the reactor.

Problem 7.5.   A. Consider the partial level schemes shown in Figure 7.3.

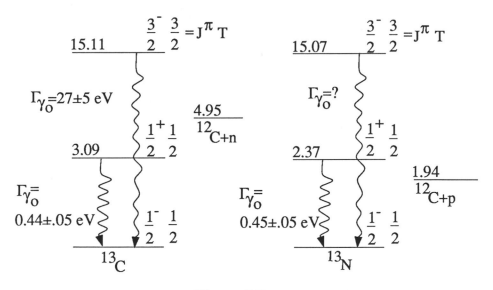

Figure 7.3.

a) Predict the Γ_{γ_0} width of the 15 MeV state in ^{13}N.

b) The $1/2^+$ states in the two nuclei appear at very different energies relative to the ground states. Why?

c) It is thus surprising that the two $1/2^+$ states have the same $\Gamma_{\gamma 0}$ width. Why? What would a naive estimate of the ratio of the widths be? What must be going on here?

B. Consider the partial level scheme shown in Figure 7.4.

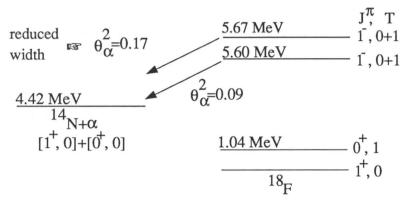

Figure 7.4.

a) Assuming two-state mixing of the 1^- levels, estimate the Coulomb mixing matrix element between them.

b) Estimate the ratio of the gamma-transition strengths of the two states to the 1.04 level, and the ratio to the ground state.

Problem 7.6. a) Give the Z and A dependence of the bulk-binding, surface, and Coulomb terms in the semi-empirical mass equation, introducing the respective constants a_V, a_S, and a_C. Neglect the pairing term throughout this question.

b) Estimate the value of a_C (in MeV).

c) The asymmetry contribution may be taken as $a_A(A - 2Z)^2/A$. *Very briefly* justify the form of this term. By considering the fact that naturally occurring $^{125}_{52}$Te is beta-stable, show $a_A/a_C \sim 30$ from the condition of minimum mass at $A = 125$.

d) The nucleus $^{264}_{100}$Fm spontaneously undergoes cold symmetric fission forming only two identical nuclei in their ground states. Assuming spherical nuclei, find the contributions of the a_V, a_S, a_C, and a_A terms to the energy release. Does fission energy primarily manifest the strong or the electromagnetic interaction? (Note: $a_V \approx a_S \approx 17$ MeV.)

Problem 7.7. The 2.15 MeV ($J^\pi = 1^+$) level in ^{10}B is populated via the ^{11}B(^{3}He,α)^{10}B reaction. It subsequently decays to the 0^+ level at 1.74 MeV via the emission of a photon with angular distribution about the beam axis of $3 + \cos^2 \theta$.

a) What is the multipolarity of the radiation?

b) The decay might also proceed via the emission of a particle of spin and parity $J^\pi = 0^-$. Deduce the form of the angular distribution of this particle with respect to the beam axis.

c) Suppose that the 2.15 MeV level is formed via the gamma decay of a higher excited 0^+ state. What would the angular correlation be between the two photons for the subsequent gamma decay to the 1.74 MeV level?

Problem 7.8. The lowest levels of the deformed nuclei dysprosium 164 and holmium 163 have been studied by reaction spectroscopy:

$$^{165}_{67}\text{Ho}_{98} \ (\text{d}, {}^3\text{He}) \ ^{164}_{66}\text{Dy}_{98} \quad \text{and} \quad ^{165}_{67}\text{Ho}_{98} \ (\text{p}, \text{t}) \ ^{163}_{67}\text{Ho}_{96} . \qquad (7.1)$$

The spin and inferred parity of stable holmium 165 is $7/2^-$. The energies, radiative lifetimes and magnetic moments observed are shown in Figure 7.5.

a) Suggest spin-parity assignments for the states observed in each nucleus. *Briefly* indicate your reasoning. Check using energy data.

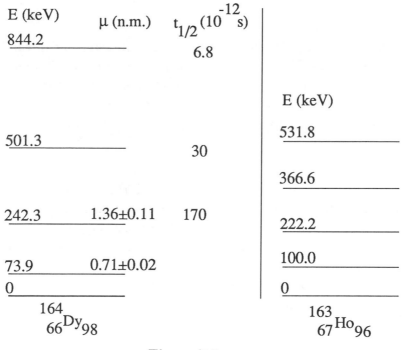

Figure 7.5.

b) Provide a vector-angular-momentum diagram for each nucleus when it is in a state of total angular momentum **J**. Include the total angular momentum **J**, its laboratory component $M\hbar$, collective angular momentum **R**, and the component $K\hbar$ of any angular momentum on the nuclear symmetry axis.

c) Compare the values of the measured magnetic moments and ratios of lifetimes with those expected from a simple collective model.

d) Develop a formula for the magnetic moment of the ground state of an odd-A nucleus, in terms of the collective and nucleonic gyromagnetic ratios g_R and g_K. (Hint: consider the vector diagram.) Then estimate the magnetic moment of the ground state of holmium 163.

Problem 7.9.   a) Assume that a single nucleon occupies the shell model state $|Nl\pi t_z jm\rangle$ where the first four quantum numbers indicate the major oscillator level, orbital angular momentum, parity, and third component of isospin. Write *single nucleon* operators corresponding to the Fermi (V) and Gamow-Teller (A) beta decay processes, then show how the respective selection rules for ΔN, Δl, $\Delta \pi$, Δj, and Δm follow.

b) The two valence nucleons of ^{6}He and ^{6}Li can be treated in the L-S scheme where $L = 0$ for both ground states. The ground state of ^{6}He ($J^\pi = 0^+$) undergoes allowed beta decay ($t_{1/2} = 0.8$ ms) to the ground state of ^{6}Li ($J^\pi = 1^+$). Explicitly write the four two-nucleon spin-isospin wavefunctions that describe the ground states of these two nuclei. Then evaluate the sum of all squared matrix elements governing this beta-decay process.

c) The $f\tau$ value for the beta decay of the neutron is approximately 1110 s while $|G_A/G_V|^2 \approx 1.50$. Using the result of part (b) (or a guesstimate thereof), predict the $f\tau$ value for ^{6}He decay.

Problem 7.10. There is a $J^\pi = 0^+$ excited state in ^{4}He (with energy= 20.1 MeV, width= 270 keV) which dominates the cross-section for ^{3}He $+ n \rightarrow t + p$ (see Figure 7.6).

a) Explain which helicity state of an unpolarized neutron beam will be more strongly absorbed by a ^{3}He target, polarized with its spin along the neutron beam axis.

b) For low energy neutrons, absorbed strongly by such a resonance, the absorption cross-section varies as 1/(velocity of the neutron). Give a simple explanation for this dependence.

c) Calculate the polarization of an initially unpolarized beam of 1 eV neutrons after passing through a sample of 6×10^{21} atoms/cm^2 of ^{3}He,

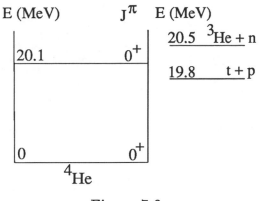

Figure 7.6.

polarized with spin along the beam axis, with a polarization of 65%. For an unpolarized beam, the absorption cross-section is $\sigma = 850$ barns at this energy and the scattering cross-section is much smaller.

Chapter 8

Elementary Particle Physics

Problem 8.1. This problem discusses K^0 decays. The quantum numbers for some mesons are:

meson	mass (MeV/c^2)	isospin	I_z	spin	parity
K^0	497.7	1/2	-1/2	0	-1
π^+	139.6	1	1	0	-1
π^0	135.0	1	0	0	-1 .

Some Clebsch-Gordon coefficients, $\langle j_1\, j_2\, m_1\, m_2 \,|\, J\, M\rangle$, are:

$$
\begin{aligned}
\langle 1\,1\,1\,-1\,|\,2\,0\rangle &= \sqrt{1/6}, & \langle 1\,1\,0\,0\,|\,2\,0\rangle &= \sqrt{2/3}, \\
\langle 1\,1\,1\,-1\,|\,1\,0\rangle &= \sqrt{1/2}, & \langle 1\,1\,0\,0\,|\,1\,0\rangle &= 0, \\
\langle 1\,1\,1\,-1\,|\,0\,0\rangle &= \sqrt{1/3}, & \langle 1\,1\,0\,0\,|\,0\,0\rangle &= -\sqrt{1/3} .
\end{aligned}
\tag{8.1}
$$

a) The short-lived neutral kaon decays quickly via $K^0 \rightarrow \pi^+\pi^-$ and $K^0 \rightarrow \pi^0\pi^0$. What values of isospin may the $\pi^+\pi^-$ and $\pi^0\pi^0$ final states assume?

b) The $K_s^0 \rightarrow \pi^+\pi^-$ decay occurs with roughly twice the probability of the $K_s^0 \rightarrow \pi^0\pi^0$ mode. Why?

c) Neutral kaons also decay via $K^0 \rightarrow \pi^+\pi^-\pi^0$ and $K^0 \rightarrow \pi^0\pi^0\pi^0$, but the three-pion decay rates are nearly three orders of magnitude lower than the two-pion rates. What accounts for the difference?

d) The CP eigenstates of the neutral K system are linear combinations of the strangeness eigenstates $| K^0 \rangle$ and $| \overline{K^0} \rangle$. If $CP| K^0 \rangle = | \overline{K^0} \rangle$, what are the CP eigenstates?

e) Let $| K_1 \rangle$ and $| K_2 \rangle$ be the CP $+1$ and CP -1 eigenstates. The K_1 component in a neutral K beam decays with rate Γ_1 and the K_2 component with rate Γ_2. If an initially pure K^0 beam is produced, what is the ratio of the number of $\overline{K^0}$ to the number of K^0 found in the beam as a function of proper time? (Assume the weak interaction is CP-conserving.)

Problem 8.2.   The scattering of high-energy muons from nucleons can provide information about the internal structure of protons and neutrons. A scattering event is described in terms of q^2, the square of the four-momentum transferred to the target, and ν, the energy lost by the muon in the collision. In the quark model, the interaction is viewed as an *elastic* collision between a muon and a stationary quark, where the quark carries a fraction x of the nucleon's mass.

a) Calculate x in terms of q^2, ν, and the nucleon mass.

b) Using simple arguments, calculate the ratio of the cross-sections for muon-proton scattering and muon-deuteron scattering, assuming $q^2 \approx 100(\mathrm{GeV}/c^2)^2$.

c) Recent experiments find a small difference in the μ^+N and μ^-N cross-sections (where N is either a proton or neutron) which grows with q^2. Explain this. What sets the scale in q^2 for this effect?

d) Measurements of nucleon structure indicate that only about half the momentum of a fast-moving nucleon is carried by objects which will scatter muons. What carries the rest of a nucleon's momentum, and why is it transparent to muons?

Problem 8.3. Scalar quarks σ are hypothetical color triplet particles with spin zero and charge $q = -e/3$. Assume their mass is 5 GeV.

a) What are the J^{PC} assignments for the low-lying σ-$\bar{\sigma}$ bound states? ($\bar{\sigma}$ is the antiparticle of σ.) Which of these could be produced in $e^+ e^-$ annihilation?

b) Take the QCD potential to be a linear confining potential $V(r) = ar$, with $a = (400 \text{ MeV})^2$ and r the σ-$\bar{\sigma}$ separation. Estimate the ground-state energy, the splitting of the low-lying states, and the value of the wavefunction squared at the origin, $|\psi(0)|^2$.

c) Write down a gauge-invariant Lagrangian describing the interaction of photons with scalar quarks. Draw the lowest-order Feynman diagrams which contribute to the decay of the σ-$\bar{\sigma}$ ground state to two photons and give a rough estimate of the width for this decay.

d) Give an argument against the existence of stable scalar quarks with mass of order a few GeV.

Problem 8.4.   The lightest charmed meson is the $D^0(1865)$ which is thought to be in a $c\bar{u}$ quark configuration. A prominent decay of this meson is the semi-leptonic transition $D \to \mu\nu X$, where X consists of hadrons.

a) Estimate the branching fraction for D^0 decays to $\mu\nu X$, and estimate the lifetime of the D^0.

b) The antiparticle of the D^0 is the $\bar{D}^0$ ($= \bar{c}u$), which is distinct, but also neutral. This allows a possible "mixing" of these two states, similar to that found in the $K^0 - \bar{K}^0$ system. However, because of the short lifetime found in (a), it is hard to demonstrate this mixing in the laboratory. One method now being pursued is as follows: prepare an initial state of pure D^0 (via a strong interaction). Then simply observe the probabilities of μ^+ and μ^- occuring in the semi-leptonic decay of the D, integrated over all time. Calculate the ratio

$$\text{Prob}(\mu^-)/\text{Prob}(\mu^+)$$

in terms of relevant parameters of the decay eigenstates of the $D^0 - \bar{D}^0$ system. You may ignore CP violation. What feature of the $D^0 - \bar{D}^0$ system contrasts with the $K^0 - \bar{K}^0$ case which makes it unlikely this ratio will be of order 1?

Problem 8.5.   The SLC (Stanford Linear Collider) is designed to produce the neutral vector boson (Z^0) by colliding electrons and positrons at an energy which can be adjusted to equal the Z^0 mass of ≈ 90 GeV. The Z^0 will manifest itself by subsequent decays to lepton or quark pairs. The vertices for the coupling of the Z^0 to all relevant particles are

$$\mathcal{L}_Z = \frac{ig}{4\cos\theta_W} Z^\mu \left\{ \bar{\nu}\gamma_\mu(1-\gamma_5)\nu + \bar{e}\gamma_\mu[(4\sin^2\theta_W - 1) - \gamma_5]e \right. \quad (8.2)$$
$$\left. + \bar{u}\gamma_\mu[(1 - \frac{8}{3}\sin^2\theta_W) + \gamma_5]u + \bar{d}\gamma_\mu[(\frac{4}{3}\sin^2\theta_W - 1) - \gamma_5]d \right\},$$

where $[\nu, e]$ and $[u, d]$ are the usual lepton and quark members of a generation. The coupling g is related to the Fermi constant G_F by $G_F = g^2\sqrt{2}/(8M_W^2)$, where $M_W \approx 80$ GeV.

a) Estimate the width $\Gamma_{\nu\bar{\nu}}$ for Z^0 decay into a given species of neutrino. Given that $\sin^2\theta_W \approx 0.23$, that there are three generations, and that the hadronic width is given adequately by the width for decay into quarks, estimate the total width of the Z^0. (In this, and the other parts of this question, you may set all relevant fermion masses equal to zero.)

b) Estimate the total cross-section for $e^+e^- \to \bar{\nu}\nu$ (one species of neutrino) via the obvious process (shown in Figure 8.1). Give the cross-

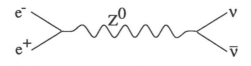

Figure 8.1.

section as a function of the square of the electron-positron center-of-mass energy, $s = E_{CM}^2$.

c) This cross-section diverges at $s = M_Z^2$ as $(s - M_Z^2)^{-2}$. This is un-physical and is properly dealt with by the Breit-Wigner trick of displacing the propagator pole off the real axis by $(s - M^2) \to (s - M^2 + iM\Gamma)$, where Γ is the total decay width of the resonance. Use this insight to improve the result of part (b) and estimate the maximum total cross-section for $e^+e^- \to anything$ as a function of energy.

d) If the SLC reaches its design-goal luminosity of $L = 10^{30} \mathrm{cm}^{-2}\mathrm{s}^{-1}$, how many Z^0 events can it produce in a year of running?

————————

Problem 8.6.   The supernova 1987A is located about 170,000 light years from earth. The interactions of 10 neutrinos from the supernova were observed in a tank of 1000 tons of water within an interval of 1 second. The average neutrino energy was 10 MeV, and the energies varied from 5 to 20 MeV. The experimental signature of each neutrino interaction included recoil nucleons as well as a charged lepton.

a) What are the weak interactions most likely involved in the creation and subsequent detection of the neutrinos?

b) Estimate an upper limit on the neutrino mass from the observed data.

c) Estimate the total energy in ergs that was liberated in the form of neutrinos during the supernova.

d) Conceivably neutrinos of type a created in the supernova could have transformed to another type b while in transit, if both types of neutrino have nonzero mass. Derive an expression for the probability that neutrinos initially of type a will appear to be type a at the Earth in terms of the mixing parameter

$$\sin \theta = \langle \nu_2 | \nu_a \rangle,$$

where ν_1 and ν_2 are neutrino states of definite mass. Deduce a limit on the mass difference between neutrinos of types 1 and 2, supposing the corresponding "oscillation length" is greater than the distance to the supernova.

Problem 8.7. The W^+ is one of the intermediate vector bosons that transmit the weak force. It was first observed at the CERN $p\bar{p}$ collider in the process

$$p\bar{p} \rightarrow W^+ + \text{hadrons}, \quad W^+ \rightarrow e^+ \nu_e. \tag{8.3}$$

The mass of the W^+ is $M_W = 82$ GeV. In the quark model, the fundamental production process is a collision between a u quark in the proton and a $\bar{d}$ antiquark in the antiproton.

If $\hat{s}$ is the square of the center-of-mass four-momentum of the u and $\bar{d}$, the cross-section for the entire process can be written in the Breit-Wigner form

$$\sigma^{u\bar{d} \rightarrow W \rightarrow e\nu}(\hat{s}) = 16\pi \frac{N_W}{N_i} \frac{\Gamma_{u\bar{d}} \Gamma_{e\nu}}{(\hat{s} - M_W^2)^2 + M_W^2 \Gamma_{tot}^2}. \tag{8.4}$$

Here N_W and N_i are the multiplicity factors for the W^+ and the initial state, Γ_{tot} is the width of the W^+, and $\Gamma_{e\nu}$ and $\Gamma_{u\bar{d}}$ are the partial widths for $W \rightarrow e\nu$ and $W \rightarrow u\bar{d}$.

a) What are the values of N_W and N_i?

b) Give an estimate of $\Gamma_{e\nu}$ in terms of G_F, the Fermi constant measured in low-energy processes such as muon decay. (Do not worry about factors of, say, 6π.) What are $\Gamma_{u\bar{d}}$ and Γ_{tot} in terms of $\Gamma_{e\nu}$? List your assumptions. What is the lifetime of the W^+ in seconds?

If $\Gamma_{tot} \ll M_W$, we can make the approximation

$$\sigma_{tot}^{u\bar{d}\to W\to e\nu} \approx \Gamma_{tot}\sigma(\hat{s} = M_W^2)M_W\delta(\hat{s} - M_W^2). \qquad (8.5)$$

(Use this even if your answer to part (b) doesn't have $\Gamma_{tot} \ll M_W$.) To find the actual cross-section, we must relate the fundamental process to one involving real protons and antiprotons. Assume that $u(x_p) = 6(1 - x_p)^2$ is the probability of a u quark carrying a fraction x_p of a proton's momentum and that $\bar{d}(x_{\bar{p}}) = 3(1 - x_{\bar{p}})^2$ is the analogous function for $\bar{d}$ antiquarks in an antiproton.

c) What is $\hat{s}$ in terms of x_p, $x_{\bar{p}}$, and s, where s is the four-momentum squared of the $p\bar{p}$ collision?

d) At CERN, this process was observed at $\sqrt{s} = 540$ GeV. What is $\sigma_{tot}(p\bar{p} \to W^+ \to e^+\nu_e)$ at this energy? The luminosity of the CERN $p\bar{p}$ collider was $L \approx 10^{29}$ cm^{-2}s^{-1}. Assuming unit efficiency, how many $p\bar{p} \to W^+ \to e^+\nu_e$ could be detected in one year of running?

Problem 8.8. In what follows assume there are six quarks with masses $m_d = m_u = m_s = m_c = 0$, $m_b = 5$ GeV, and $m_t = 40$ GeV. Selected elements of the Kobayashi-Maskawa mixing matrix along with various other experimental data are given below.

$$\begin{bmatrix} d' \\ s' \\ b' \end{bmatrix} = \begin{bmatrix} V_{ud} & V_{us} & V_{ub} \\ V_{cd} & V_{cs} & V_{cb} \\ V_{td} & V_{ts} & V_{tb} \end{bmatrix} \begin{bmatrix} d \\ s \\ b \end{bmatrix}$$

$$= \begin{bmatrix} 0.973 \pm 0.005 & 0.221 \pm 0.003 & V_{ub} \\ 0.207 \pm 0.024 & 0.66 \text{ to } 0.98 & V_{cb} \\ V_{td} & V_{ts} & V_{tb} \end{bmatrix} \begin{bmatrix} d \\ s \\ b \end{bmatrix}$$

$$m_\mu = 106 \text{ MeV}$$

$$\tau_\mu = 2.20 \times 10^{-6}\text{s}$$

$$BR(B^\pm \rightarrow e^\pm \nu + \text{hadrons}) = 0.12$$

$$\frac{\Gamma(b \rightarrow u)}{\Gamma(b \rightarrow c)} < 0.08$$

$$\tau_{B^\pm} = 1.4 \times 10^{-12}\text{s}$$

a) Estimate $|V_{cb}|$.

b) Use your result from part (a) to estimate the largest allowable value of $|V_{ub}|$.

c) Estimate the top-quark lifetime.

Problem 8.9. Because of its role in giving mass to fermions, the still-hypothetical Higgs scalar couples to each fermion with a strength proportional to the fermion's mass (Figure 8.2).

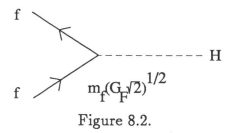

Figure 8.2.

a) Calculate the decay width $\Gamma(H \rightarrow f\bar{f})$, where $f\bar{f}$ is a fermion-antifermion pair with $m_f \ll m_H$, m_f is the fermion mass and m_H is the Higgs mass. What is the lifetime of a 50 MeV Higgs decaying to an electron-positron pair?

b) Consider the reaction $e^+ e^- \rightarrow H \rightarrow f\bar{f}$. For a 50 GeV Higgs, $f\bar{f}$ is predominantly bottom-antibottom ($m_b = 5$ GeV). What is the cross-section for this reaction at resonance? How many events of this type would be produced at an electron-positron collider, running for a year at the resonance, with a luminosity of 2×10^{31} cm^{-2}sec^{-1}?

Problem 8.10.   A high-energy neutrino experiment scatters muon-neutrinos off a fixed target of nuclei containing equal numbers of protons and neutrons. The measured parameter is R_ν, the ratio of the neutral-current to   charged-current total cross-sections,

$$R_\nu = \frac{\sigma_{NC}}{\sigma_{CC}} = 0.3 \,. \tag{8.6}$$

From this result the value of the Weinberg angle which governs the relation between the electromagnetic and weak neutral-current interactions can be calculated.

(a) Draw the Feynman diagram(s) for the fundamental charged- and neutral-current interactions in this experiment. Assume that nucleons contain only up and down quarks.

The differential cross-sections for neutrino-quark scattering are

$$\frac{d\sigma}{dy} = \begin{cases} \dfrac{G_F^2\, s}{\pi}\overline{Q_W^2}\,, & \text{target left-handed,} \\[4mm] \dfrac{G_F^2\, s}{\pi}(1-y)^2\overline{Q_W^2}\,, & \text{target right-handed,} \end{cases} \tag{8.7}$$

where s is the square of the center of mass energy, y is the fraction of the neutrino's energy transferred to hadrons, G_F is the Fermi constant, and Q_W is the "weak charge" of the target quark. The bar over Q_W^2 refers to the fact that this quantity must be averaged over the quarks in the target. In the Weinberg-Salam electroweak theory, the weak charges

$$Q_W = L_{CC}, R_{CC}, L_{NC}, R_{NC}, \tag{8.8}$$

are given as follows:

	left-handed fermion	right-handed fermion
charged current	$L_{CC} = 1$	$R_{CC} = 0$
neutral current	$L_{NC} = I_3 - Q_e \sin^2 \theta_W$	$R_{NC} = I_3 - Q_e \sin^2 \theta_W$

where I_3 is the third component of weak isospin, Q_e is the electric charge of the quark measured in units of the absolute value of the electron charge, and θ_W is the Weinberg angle.

(b) Use this information and the measured value of R_ν given above to calculate the Weinberg angle. Again assume only up and down quarks and neglect possible differences in momentum distributions between the quarks.

Chapter 9

Atomic & General Physics

Problem 9.1. Positronium is a hydrogen-like bound state made up of an electron and a positron.

a) Estimate the binding energy of the ground state ($n = 1$) and the Lyman-alpha ($2p \rightarrow 1s$) transition wavelength for positronium.

b) The lifetime for decay from $2p$ to $1s$ for the hydrogen atom is 1.6 ns. Estimate the lifetime for the same decay in positronium.

c) Estimate the strength of the magnetic field experienced by the electron due to the positron's magnetic moment for the $n = 1$ state.

d) Estimate the singlet-triplet frequency splitting in the ground state.

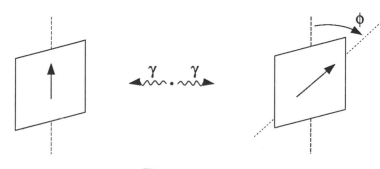

Figure 9.1.

e) The singlet state decays into two 0.511 MeV γ rays in a state of definite parity. If the γ rays are detected in coincidence after passing through linear polarizers (as in Figure 9.1), at what relative angle ϕ is a maximum coincidence rate expected?

Problem 9.2.   Consider a lens of focal length f with a diameter

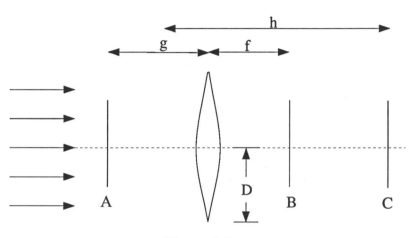

Figure 9.2.

D (Figure 9.2). A laser beam of wavelength λ illuminates the entire aperture of the lens. The laser beam is parallel to the axis of the lens.

a) Suppose a screen is placed at B. Approximately how wide is the image at B?

b) Now place at A a transmission grating with line separation d. Describe the pattern at B.

c) Suppose you wish to construct a microscope to photograph a microbe, which is placed at A (in the plane of the grating). The film is placed at C. How far is C from the lens?

d) Since many microbes are transparent, a "phase plate" greatly improves the microscope. Light that goes through the circular phase plate within x of the center suffers a $\pi/2$ phase lag. The plate is put at B, centered on the optical axis of the lens, with its surface normal to the optical axis. Suppose the index of refraction of the microbe differs slightly from that of the surrounding medium so that light which passes through the microbe is slightly phase-shifted. Compare the image with the addition of the phase plate to the image without the phase plate. How large should x be? If the center of the phase plate is opaque, how is the image changed?

Problem 9.3.   a) A plane wave is incident on a lens of index of refraction n with one flat side (as shown in Figure 9.3). The diameter of the lens is D. If all incoming rays are to be focused at the point P a distance f from the flat side, calculate the shape of the curved side. (Do not assume the lens is thin.)

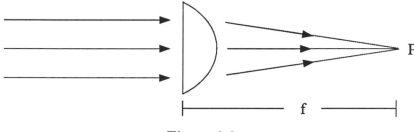

Figure 9.3.

b) Where is the focal point of two parallel thin lenses with focal lengths f_1 and f_2 if their axes are slightly misaligned by a distance δ? Give both horizontal and vertical position. The lenses are separated by a distance D where $D < f_1, f_2$ (see Figure 9.4).

Problem 9.4.   A cloud of neutral hydrogen atoms in interstellar space

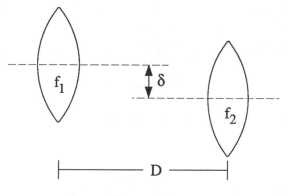

Figure 9.4.

has density ρ and temperature T. Write an approximate expression (drop factors of 2 but retain factors of the fine structure constant α, $\hbar$, T, etc.) for:

a) the frequency of the absorption line due to transitions from the ground state to a $2p$ level. Ignore fine structure,

b) the natural width of this line,

c) the Doppler width,

d) the linewidth due to collisions,

e) the fine structure splitting of the $2p$ level, and

f) the hyperfine splitting of the $1s$ level.

Problem 9.5. The magnetic hyperfine hamiltonian is $H_{hyp} = A\hbar\mathbf{I} \cdot \mathbf{J}$ where A is a constant, $\mathbf{I}$ is the nuclear angular momentum, and $\mathbf{J}$ is the total electronic angular momentum. Consider an atom with electronic ground state g and an excited state e. In the presence of the hyperfine interaction each of these levels is split.

States g and e both have electronic angular momentum $J = 1/2$, and the nucleus has spin $I = 1/2$. At time $t = 0$, an infinitesimally short laser pulse excites atoms from the ground state g to the excited state e. The light from the laser is circularly polarized and its linewidth is much greater than the hyperfine splitting. Taking the laser-propagation direction to be the quantization axis, the excited-state wavefunction immediately after the laser pulse is given by

$$\psi(0) = \frac{1}{\sqrt{2}} \left| J_e = \tfrac{1}{2}, m_J = \tfrac{1}{2} \right\rangle \left\{ \left| I = \tfrac{1}{2}, m_I = -\tfrac{1}{2} \right\rangle + \left| I = \tfrac{1}{2}, m_I = \tfrac{1}{2} \right\rangle \right\}.$$
(9.1)

The atoms then decay from the excited state to the ground state with lifetime τ. Calculate the time dependence of the intensity of circularly polarized fluorescence having the same handedness as the exciting laser pulse.

Problem 9.6. Estimate the following quantities, indicating how you arrived at your estimate:

a) the frequency of radiation used in a microwave oven,

b) the energy yield of a fission bomb with a 30 kg uranium core,

c) the energy of impact of the earth with a meteorite 10 meters in diameter,

d) the temperature of the sun,

e) the temperature of a 60-watt light bulb filament,

f) the speed of sound in helium gas at STP,

g) the total length of blood capillaries in the human body, and

h) the temperature at which the heat capacity for gaseous molecular hydrogen changes from $3k/2$ to $5k/2$ per molecule.

Problem 9.7. a) List the ground-state electronic configurations and the L, S, and J quantum numbers for the following atoms:

$$\text{Li}(Z=3), \quad \text{B}(Z=5), \quad \text{N}(Z=7), \quad \text{Na}(Z=11), \quad \text{K}(Z=19) \, .$$

b) The lowest frequency line in the absorption spectrum of Na is a doublet. What mechanism splits the corresponding pair of energy levels? The splitting between levels is proportional to $\langle r^n \rangle$ where r is the distance of the valence electron from the nucleus. What is a numerical value for n?

c) Consider the effect of a weak magnetic field B on the low-lying states of potassium $(Z=19)$. Make a sketch indicating the allowed transitions and give the energy splitting between states as a function of B.

Problem 9.8. A spherical projectile of radius R and density μ moves through a fluid of density ρ and viscosity η at velocity v.

a) For very small velocities what is the frictional force (drag) on the projectile? Throughout this problem, you may set pure numerical factors equal to unity.

b) For high velocities, the drag is independent of the viscosity. Write an expression for the drag in this regime. Interpret your result in terms of the acceleration of the fluid. What is the characteristic crossover velocity v_c between the low and high velocity regimes?

c) If the initial velocity is $v_0 \gg v_c$, how far will the projectile travel before stopping?

Problem 9.9. Consider a tuned infrared coherent light source ($\lambda \approx 1$ to 10 μm) which utilizes stimulated Raman scattering in strontium atoms (atomic number 38). Here the atom absorbs a pump photon of

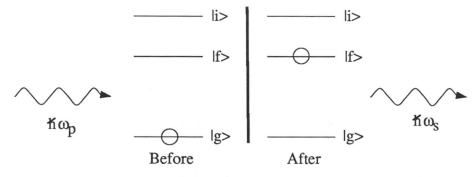

Figure 9.5.

energy $\hbar\omega_p$ (visible light) and undergoes a virtual transition from the ground state $|g\rangle$ to a state $|i\rangle$ (Figure 9.5). This results in the emission of a signal photon of energy $\hbar\omega_s$ and leaves the atom in the state $|f\rangle$.

a) Assume that the signal intensity $I_s(0)$ at the input end of an atomic vapor cell of length l is much less than the intensity of the pump beam, $I_p(0)$. Let the pump beam be incident along the $+\hat{z}$-axis. The growth of the original intensity $I_s(z)$ is given by

$$\frac{d}{dz}I_s(z) = gI_p(z)I_s(z) , \tag{9.2}$$

where $I_p(z)$ is the pump intensity and g is the Raman gain. Find an expression for $I_s(z)$ in terms of $I_s(0)$ and $I_p(0)$ and show that the expression has the correct form in the small signal limit, i.e., in the case where most of the atoms are still in the ground state after the pump beam passes through.

b) Explain how the signal radiation is tunable.

c) What are the parities of the states $|g\rangle$, $|i\rangle$, and $|f\rangle$? What are the appropriate states for strontium?

Problem 9.10. Hurricane Gilbert, when viewed from space, was an enormous spiral of clouds surrounding a small, cloud-free "eye." Use basic physics and any other knowledge you have about hurricanes in the northern hemisphere to answer the following questions.

a) What was the direction of rotation of the spiral and why?

b) The minimum atmospheric pressure at the eye of Gilbert was only 86% of normal atmospheric pressure at sea level. Use this pressure reduction to estimate the wind velocity near the eye of the hurricane. You may approximate the air as incompressible and ignore any cooling or heating of the air.

c) On a dry, sunny day in Princeton the air temperature is found to drop with increasing altitude above ground level by about 10°C/km. Show how to predict this temperature drop.

d) The air temperature near the eye of Hurricane Gilbert was observed to drop with increasing altitude above sea level at a rate of only 4°C/km. Explain why this temperature gradient differs from that of part (c).

Possibly useful information:
Atmospheric pressure at sea level: 10^6 dyne/cm^2
Air density at sea level: 1.275 mg/cm^3
Acceleration of gravity at sea level: 981 cm/sec^2
Apparent molecular weight of air: 29 g
Specific heat of air at constant pressure: $c_p = 10^7$ erg °C^{-1}g^{-1}
Latent heat of vaporization of water: $L = 2.3 \times 10^{10}$ erg/g
Saturation vapor pressure of water at 20 °C: 2×10^4 dyne/cm^2
Boltzmann's constant: 1.38×10^{-16} erg °C^{-1}

Part II

Solutions

Chapter 10

Mechanics—Solutions

Solution 1.1. First we set up coordinates such that the ball moves in the xy-plane with $\hat{y}$ directed upward and $\hat{z}$ out of the page. When the ball bounces, it experiences a tangential force F_x which changes its momentum in the x-direction:

$$\Delta p_x = \int F_x \, dt = m(v_f - v_i), \qquad (10.1)$$

where v_i and v_f are the x-components of the velocity of the center of mass of the ball before and after the bounce. The force also exerts a torque on the ball. If a is the radius of the ball, then the change in its angular momentum $\Delta \mathbf{L}$ is

$$\Delta \mathbf{L} = a \int F_x \, dt \, \hat{z} = I(\omega_f - \omega_i)\hat{z}, \qquad (10.2)$$

where $I = 2ma^2/5$ is the moment of inertia of the Super-Ball, and ω_i and ω_f are its angular frequences before and after the bounce. Symmetry requires that the ball spin about an axis parallel to $\hat{z}$. Elimination of the unknown force integral between these equations leaves

$$(v_f - v_i) = \frac{2}{5}a(\omega_f - \omega_i). \qquad (10.3)$$

For the ball to bounce in the prescribed manner, v_f must equal $-v_i$. From symmetry arguments, we must also have $\omega_f = -\omega_i$. Using these

77

two conditions in the above equation yields the requirements on the original throw:

$$v_i = \frac{2}{5} a \omega_i. \tag{10.4}$$

We may also derive the condition $|\omega_f| = |\omega_i|$ from conservation of energy. For completeness, this derivation follows. Because the collision is elastic, energy is conserved. The magnitude of the momentum in the y-direction is unchanged by the collision, so the energy balance equation becomes

$$\frac{1}{2} m v_i^2 + I \omega_i^2 = \frac{1}{2} m v_f^2 + I \omega_f^2. \tag{10.5}$$

Since $|v_i| = |v_f|$, it must be that $|\omega_i| = |\omega_f|$. From equation (10.3) we can see that $v_i = -v_f$ requires $\omega_i = -\omega_f$.

Solution 1.2. Suppose the rocket is moving in the positive x-direction, and the dust cloud starts at $z = 0$. Because the collisions between the rocket and dust particles are inelastic, energy is not conserved. However, we must conserve momentum at all times. If $m(x)$ and $v(x)$ are the mass and velocity of the rocket at point x, then for all x

$$m(x)v(x) = m_0 v_0. \tag{10.6}$$

In particular, for a small displacement δx,

$$m(x)v(x) = m(x + \delta x)v(x + \delta x). \tag{10.7}$$

As the rocket travels from the edge of the cloud to a point x, it sweeps the dust out of a region of volume Ax, so its mass at position x is

$$m(x) = m_0 + A\rho x. \tag{10.8}$$

Expanding equation (10.7) gives us

$$m(x)v(x) = (m(x) + A\rho\delta x)\left(v(x) + \frac{dv}{dx}\delta x + \mathcal{O}(\delta x^2)\right). \tag{10.9}$$

Neglecting terms of second and higher order in δx,

$$\left[A\rho v(x) + m(x)\frac{dv}{dx}\right] = 0, \tag{10.10}$$

or, using (10.8),

$$\frac{A\rho}{m_0 + A\rho x}\,dx + \frac{dv}{v} = 0. \tag{10.11}$$

Integrating this equation and using the initial condition $v(x = 0) = v_0$, we find that

$$v = \frac{dx}{dt} = \frac{m_0 v_0}{m_0 + A\rho x}. \tag{10.12}$$

Integrating again and using the condition that $x(t = 0) = 0$, it is easy to show that, for $t > 0$,

$$x(t) = -\frac{m_0}{A\rho} + \sqrt{\frac{2m_0 v_0 t}{A\rho} + \frac{m_0^2}{A^2 \rho^2}}. \tag{10.13}$$

Solution 1.3. In order for the "satellite" to remain in orbit, the centripetal acceleration due to its orbital motion must exactly balance the gravitational acceleration, so

$$\int_R^{R+l} r\omega^2 \rho\,dr = \int_R^{R+l} \frac{GM}{r^2}\rho\,dr, \tag{10.14}$$

where ρ is the density of the rope and l is its length. Performing the integration yields

$$\frac{2GM}{\omega^2} = R(R+l)(2R+l). \tag{10.15}$$

Solving this quadratic equation gives us the length of the rope

$$l = \frac{-3R + \sqrt{9R^2 + 4\left(2GM/R\omega^2 - 2R^2\right)}}{2}. \tag{10.16}$$

To find a numerical value for l we must recall certain physical constants. The radius of the earth is $R = 6.4 \times 10^6$ m. (One way to remember this is to recall that the meter was first defined as one ten-thousandth the distance from the North Pole to the equator.) The rate of rotation of the earth is $\omega = 2\pi$ rad/24 hours $= 7.3 \times 10^{-5}$rad/s. Noticing that $GM/R^2 = g = 9.8$ m/s^2, we avoid having to know the values of G and M separately. Using these numbers, we calculate $l = 1.5 \times 10^8$m (or almost halfway to the moon!).

Solution 1.4. We will use the normal modes of the system to solve this problem. First we find the motion of the normal-mode coordinates subject to the applied force, and then transform from those coordinates to the ordinary spatial coordinates of the blocks. With some physics insight we could immediately write down the normal modes of the system (10.27). However, it is instructive to solve methodically for the normal modes, as we do below.

Let η_i be the displacement of block i from equilibrium. The potential energy of the system is

$$V = \frac{1}{2}k(\eta_1 - \eta_2)^2 + \frac{1}{2}k(\eta_2 - \eta_3)^2, \qquad (10.17)$$

and the kinetic energy is

$$T = \frac{1}{2}m(\dot{\eta_1}^2 + \dot{\eta_2}^2 + \dot{\eta_3}^2). \qquad (10.18)$$

The Lagrangian is $L = (T - V)$, which we can write as

$$L = \frac{1}{2}\sum_{i=1}^{3}\sum_{j=1}^{3}(T_{ij}\dot{\eta_i}\dot{\eta_j} - V_{ij}\eta_i\eta_j), \qquad (10.19)$$

where

$$\mathbf{T} = \begin{pmatrix} m & 0 & 0 \\ 0 & m & 0 \\ 0 & 0 & m \end{pmatrix} \text{ and } \mathbf{V} = \begin{pmatrix} k & -k & 0 \\ -k & 2k & -k \\ 0 & -k & k \end{pmatrix}. \qquad (10.20)$$

For our simple problem it may seem that writing $\mathbf{T}$ as a matrix is unnecessary and heavy-handed. However, it is useful to do this so that we could easily generalize to the case in which the masses of the particles are not equal.

Using Lagrange's equation, we find the equations of motion:

$$\mathbf{T}\ddot{\eta} + \mathbf{V}\eta = 0, \tag{10.21}$$

where we have defined the vector $\eta = (\eta_1, \eta_2, \eta_3)^T$. The normal modes are collective motions where all three blocks move with the same frequency. Since there are three degrees of freedom there will be three normal modes. For each one the solution is of the form

$$\eta(t) = \mathbf{a}_j e^{i\omega_j t}, \tag{10.22}$$

where the $\mathbf{a}_j$ are time-independent. If we insert this form for $\eta(t)$ into the equations of motion (10.21), we get a matrix equation for the vector $\mathbf{a}_j$,

$$(\mathbf{V} - \omega_j^2 \mathbf{T})\mathbf{a}_j = 0. \tag{10.23}$$

In order for a nontrivial solution to exist, we must have

$$\det\left[\mathbf{V} - \omega_j^2 \mathbf{T}\right] = 0. \tag{10.24}$$

This leads to a cubic equation in ω^2, with roots $\omega_1^2 = 0$, $\omega_2^2 = k/m$, and $\omega_3^2 = 3k/m$. Substituting these frequencies into equation (10.23) allows us to solve for the three normal modes, for which we choose the normalization prescription

$$\mathbf{a}_i^T \mathbf{T}\, \mathbf{a}_i = 1 \quad \text{(no summation on i)}. \tag{10.25}$$

In fact it can be shown (see Goldstein, chapter 6) that the vectors $\mathbf{a}_i$ may be chosen to satisfy the "orthogonality" condition

$$\mathbf{a}_i^T \mathbf{T}\, \mathbf{a}_j = \delta_{ij}, \tag{10.26}$$

and we will use this later on. Subject to this condition, our normal modes are

$$\mathbf{a}_1 = \frac{1}{\sqrt{3m}} \begin{pmatrix} 1 \\ 1 \\ 1 \end{pmatrix}, \quad \mathbf{a}_2 = \frac{1}{\sqrt{2m}} \begin{pmatrix} 1 \\ 0 \\ -1 \end{pmatrix}, \quad \text{and } \mathbf{a}_3 = \frac{1}{\sqrt{6m}} \begin{pmatrix} 1 \\ -2 \\ 1 \end{pmatrix}. \tag{10.27}$$

We can use these vectors as a basis set to write an arbitrary displacement as

$$\boldsymbol{\eta}(t) = \xi_1 \mathbf{a}_1 + \xi_2 \mathbf{a}_2 + \xi_3 \mathbf{a}_3, \tag{10.28}$$

where the ξ_i are called normal coordinates.

Suppose we now apply a force $\mathbf{F}(t)$. Our equations of motion are

$$\mathbf{T} \left(\sum_{i=1}^{3} \ddot{\xi}_i \mathbf{a}_i \right) + \mathbf{V} \left(\sum_{i=1}^{3} \xi_i \mathbf{a}_i \right) = \mathbf{F}(t). \tag{10.29}$$

We use the matrix equation for a normal mode vector, (10.23), to rewrite $\mathbf{V}\mathbf{a}_i$ as $\omega_i^2 \mathbf{T} \mathbf{a}_i$. If we now multiply on the left by $\mathbf{a}_j^T$ and use the orthogonality condition (10.26), the normal modes decouple (which is why they are called normal modes) and we obtain the equations of motion for the normal coordinates:

$$\ddot{\xi}_j + \omega_j^2 \xi_j = f_j(t), \tag{10.30}$$

where we have defined

$$f_j(t) = \mathbf{a}_j^T \mathbf{F}(t). \tag{10.31}$$

In our particular problem, the force is given by

$$\mathbf{F} = \begin{pmatrix} f \cos \omega t \\ 0 \\ 0 \end{pmatrix}. \tag{10.32}$$

This gives us

$$f_1 = \frac{1}{\sqrt{3m}} f \cos \omega t, \tag{10.33}$$

$$f_2 = \frac{1}{\sqrt{2m}} f \cos \omega t, \tag{10.34}$$

$$f_3 = \frac{1}{\sqrt{6m}} f \cos \omega t. \tag{10.35}$$

It is now straightforward to solve the equations of motion (10.30), subject to the initial conditions

$$\dot{\xi}_i = 0, \quad \xi_i = 0. \tag{10.36}$$

The solution is

$$\xi_1 = \frac{f}{\sqrt{3m}\,\omega^2}(1 - \cos\omega t), \tag{10.37}$$

$$\xi_2 = \frac{f}{\sqrt{2m}(\omega_2^2 - \omega^2)}(\cos\omega t - \cos\omega_2 t), \tag{10.38}$$

$$\xi_3 = \frac{f}{\sqrt{6m}(\omega_3^2 - \omega^2)}(\cos\omega t - \cos\omega_3 t). \tag{10.39}$$

(This can be verified by substitution.) Next we substitute the normal coordinates back into equation (10.28), to find the motion of mass C:

$$\eta_3 = \frac{f}{m}\left[\frac{1}{3\omega^2}(1 - \cos\omega t) - \frac{1}{2(\omega_2^2 - \omega^2)}(\cos\omega t - \cos\omega_2 t)\right.$$
$$\left. + \frac{1}{6(\omega_3^2 - \omega^2)}(\cos\omega t - \cos\omega_3 t)\right]. \tag{10.40}$$

Solution 1.5. There are two solutions. The first is that the ball stays in one place, rolling such that $\omega = 0$. We'll look for a more interesting solution.

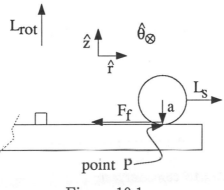

Figure 10.1.

The geometry of the problem is shown in Figure 10.1. In order for the ball to roll such that $\omega \neq 0$, there must be a radial force that will cause the centripetal acceleration. This is supplied by the static friction force $F_f \leq \mu N$. Note that this is the force that keeps the ball from sliding radially outward, and is unrelated to the lack of rolling friction. So

$$\mathbf{F}_f = -m\omega^2 r\hat{\mathbf{r}}, \qquad (10.41)$$

where m is the mass of the ball, and where we have chosen a cylindrical coordinate system centered on the turntable pivot, with $+\hat{\mathbf{z}}$ pointing out of the plane of the turntable.

The ball undergoes two motions: a spinning about its center of mass, and a rotation about the turntable pivot. The spinning motion has an angular momentum $\mathbf{L}_s$ which points in the $\pm\hat{\mathbf{r}}$ direction. The angular momentum associated with the rotation, $\mathbf{L}_{rot}$, is in the $\pm\hat{\mathbf{z}}$ direction. (The total angular momentum is $\mathbf{L}_{tot} = \mathbf{L}_{rot} + \mathbf{L}_s$, of course.)

In order for $\mathbf{L}_s$ to remain radial as the ball rotates about the pivot, $\mathbf{L}_s$ must precess. That is, the vector $\mathbf{L}_s$ rotates around the turntable with the ball. There must be a torque:

$$\tau = \frac{d\mathbf{L}_s}{dt} = \dot{\mathbf{L}}_s. \qquad (10.42)$$

The torque is provided by the same friction force responsible for the centripetal acceleration:

$$\tau = \mathbf{a} \times \mathbf{F}_f = (-a\hat{\mathbf{z}}) \times (-m\omega^2 r\hat{\mathbf{r}}) = am\omega^2 r\hat{\theta} = \dot{\mathbf{L}}_s, \qquad (10.43)$$

where $\mathbf{a}$ is the vector extending from the center of the ball to its instantaneous point of contact with the turntable. If the ball spins about its center of mass with angular velocity ω_s then $|\mathbf{L}_s| = I\omega_s$, where I is the moment of inertia of the ball about its center of mass. The precession of $\mathbf{L}_s$ is described by

$$\dot{\mathbf{L}}_s = I\omega_s\dot{\hat{\mathbf{r}}} = I\omega_s\omega\hat{\theta}. \qquad (10.44)$$

We have two cases to consider:

1) Ω and ω are in the same direction. Let us arbitrarily choose the $+z$-direction for both, or a counterclockwise motion when viewed from above. Because $\dot{\mathbf{L}}_s$ is in the $+\theta$-direction, $\mathbf{L}_s$ must be in the $+r$-direction if ω is a counterclockwise motion. (Think this one through!) This

implies that $\omega < \Omega$, because the ball must "roll backwards." (That is, the motion described by $\omega - \Omega$ is directed opposite to the motion of the turntable alone — which is described by Ω.)

The condition for the ball to roll without slipping is

$$a\omega_s = r(\Omega - \omega).\tag{10.45}$$

Let P be the point on the ball in instantaneous contact with the turntable. The above condition comes from requiring P's instantaneous speed due to the ball's rotation about the turntable, which is $r(\Omega - \omega)$, to equal the instantaneous speed due to the ball's spinning, which is $a\omega_s$.

Since the moment of inertia of the ball is $I = 2ma^2/5$, we can combine the torque equations (10.43) and (10.44) to find

$$\omega = \frac{2a\omega_s}{5r}.\tag{10.46}$$

Then, using the condition (10.45), we find

$$\omega = \frac{2}{7}\Omega.\tag{10.47}$$

2) The second case is ω and Ω in opposite directions. Our intuition suggests that this will not happen. We can demonstrate that it does not by observing that for ω clockwise and Ω counterclockwise, $\mathbf{L_s}$ must be in the $+r$-direction. However, by equation (10.43), $\dot{\mathbf{L}}_s$ is in the $+\theta$-direction, or counterclockwise, which is inconsistent with ω clockwise.

———————

Solution 1.6. a) Suppose the platform is initially at $x = 0$, where x is measured upwards. Assume the dashpot provides a damping force given by $F_{damp} = -\gamma\dot{x}$ (where $\gamma > 0$). Then the equation of motion of the platform is

$$m\ddot{x} + \gamma\dot{x} + kx + mg = 0.\tag{10.48}$$

If we make a change of coordinates to $y = x + mg/k$ and write $\gamma = 2m\alpha$, this simplifies to

$$\ddot{y} + 2\alpha\dot{y} + \frac{k}{m}y = 0. \qquad (10.49)$$

One can easily verify that the general solution to this equation is

$$y(t) = Ae^{\omega_+ t} + Be^{\omega_- t}, \qquad (10.50)$$

where $\omega_\pm = -\alpha \pm \sqrt{\alpha^2 - k/m}$. Depending on the value of k/m, $\omega_\pm$ may be complex.

Critical damping occurs when the expression in the radical vanishes, when $\alpha^2 = k/m$. In this case the two values of ω coincide and the solution is of a different form:

$$y = (A + Bt)e^{-\alpha t}. \qquad (10.51)$$

If we start measuring time from the moment the putty hits the platform, then the initial conditions are

$$y(0) = A = \frac{mg}{k} = \frac{g}{\alpha^2}, \qquad (10.52)$$

and

$$\dot{y}(0) = (B - \alpha A) = -\sqrt{2gh}. \qquad (10.53)$$

We eliminate the constants A and B to find

$$y(t) = \left[\frac{g}{\alpha^2} + \left(\frac{g}{\alpha} - \sqrt{2gh}\right)t\right]e^{-\alpha t}. \qquad (10.54)$$

Note that the sign of B, the coefficient of t in equation (10.51), determines whether there will be overshoot. If B is positive (i.e., $k/m < g/2h$) then there is no overshoot; however, if B is negative then for large t, $y(t)$ will also be negative, so overshoot occurs (Figure 10.2).

b) Suppose now the system is overdamped, i.e. $\alpha^2 > k/m$, and write

$$\omega_\pm = -\alpha \pm \sqrt{\alpha^2 - k/m}. \qquad (10.55)$$

We note that $|\omega_-| > \alpha > |\omega_+|$ and that $\omega_\pm$ are real and negative. The general overdamped solution,

$$y(t) = Ae^{\omega_+ t} + Be^{\omega_- t}, \qquad (10.56)$$

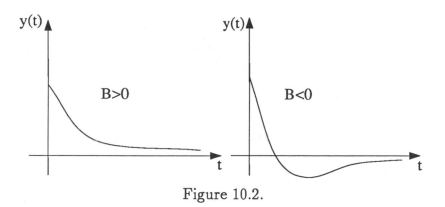

Figure 10.2.

is the sum of two decaying exponentials, where the ω_- term decays more rapidly than (and the ω_+ term more slowly than) the critical damping solution. In general the initial conditions will be such that both terms are present, in which case the ω_+ term will dominate for large t. It is for this reason that it is conventional to state in textbooks that critical damping is the "best" solution in the sense of offering the fastest approach to equilibrium. However we see that if we fine-tune the initial conditions and the damping constant γ, we can arrange for the ω_+ term to be absent. If we do this, we have a solution that decays *faster* than the critically-damped solution. Note also that since our solution is a single decaying exponential, there is no possibility of overshoot.

We feel obliged to point out that a solution of this type would seem to be of little practical use: for a real system (such as a car suspension) neither the initial conditions (the size of a pothole) nor the mass (of the loaded car) are predictable quantities.

Now we will find the condition on the amount of damping (i.e., on the size of γ) in order for the ω_+ term to vanish. The initial conditions are

$$y(0) = A + B = \frac{mg}{k}, \quad \text{and} \tag{10.57}$$

$$\dot{y}(0) = (A\omega_+ + B\omega_-) = -\sqrt{2gh}. \tag{10.58}$$

We could invert these equations and find the condition for A to vanish, or we could take a short-cut and set A to zero. After eliminating B we obtain

$$\omega_- = -\sqrt{2gh}\frac{k}{mg}. \tag{10.59}$$

Using the expression for ω_- in terms of α (10.55), we can rearrange to obtain

$$\sqrt{\alpha^2 - \frac{k}{m}} = \frac{k}{m}\sqrt{\frac{2h}{g}} - \alpha. \tag{10.60}$$

After squaring both sides and simplifying, we reach our final result:

$$\gamma = 2m\alpha = m\sqrt{\frac{g}{2h}} + k\sqrt{\frac{2h}{g}}. \tag{10.61}$$

Solution 1.7. (This argument is based on one by Landau and Lifshitz in *Mechanics*, Chapter 7.) If p and q are the momentum and position of the mass, the quantity

$$I = \oint \frac{p\,dq}{2\pi}, \tag{10.62}$$

with the integral taken over a single oscillation, is an adiabatic invariant. This means that I remains unchanged as the spring constant slowly decreases. Stoke's theorem allows us to rewrite this line integral as an integral over the area in phase space enclosed by the path followed by the mass over the course of one oscillation:

$$I = \iint \frac{dp\,dq}{2\pi}. \tag{10.63}$$

For a simple harmonic oscillator,

$$\frac{1}{2m}p^2 + \frac{1}{2}m\omega^2 q^2 = E, \tag{10.64}$$

where E is the energy of the system and $\omega = \sqrt{k/m}$. The path followed by the mass is an ellipse in phase space, with area $2\pi E/\omega$, so the adiabatic invariant is $I = E/\omega$. (This important result is true for any harmonic oscillator, and should be remembered.) The energy of a simple harmonic oscillator is related to its amplitude by

$$\frac{1}{2}m\omega^2 A^2 = E, \tag{10.65}$$

so $E_1/\omega_1 = E_2/\omega_2$ implies that

$$\frac{m\omega_1^2 A_1^2}{2\omega_1} = \frac{m\omega_2^2 A_2^2}{2\omega_2}, \tag{10.66}$$

or

$$A_2 = A_1 \left(\frac{k_1}{k_2}\right)^{1/4}. \tag{10.67}$$

In quantum mechanics, the condition that $I = E/\omega$ is constant arises naturally from the condition that the system remains in an energy eigenstate (with the same quantum numbers) during an adiabatic change. For a simple harmonic oscillator, $E = (n + \frac{1}{2})\hbar\omega$, so $E/\omega = (n + \frac{1}{2})\hbar$ is conserved.

Solution 1.8. When a film of soap is stretched across a frame, surface tension forces the film to adjust itself so that its surface area is a minimum (Figure 10.3). Hence our problem can be formulated in terms of the calculus of variations.

First we note that since the problem is axially symmetric about the line AB, we can describe the surface by specifying its radius as a function of distance along the symmetry axis, $r = f(z)$. For a given function f, the surface area is given by

$$S = \int_{z=0}^{d} 2\pi f \sqrt{df^2 + dz^2} = \int_0^d 2\pi f \sqrt{1 + f'^2}\, dz, \tag{10.68}$$

where $f' \equiv df/dz$.

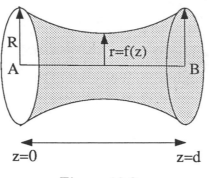

Figure 10.3.

In any text on the calculus of variations (e.g., Boas), it is shown that the function $G(z, f, f')$ which extremizes the integral

$$I = \int G \, dz \tag{10.69}$$

must satisfy the Euler-Lagrange equation:

$$\frac{d}{dz}\left(\frac{\partial G}{\partial f'}\right) - \frac{\partial G}{\partial f} = 0. \tag{10.70}$$

In this problem $G(z, f, f') = 2\pi f \sqrt{1 + f'^2}$.

At this point we could insert G into the Euler-Lagrange equation and attempt to solve the resulting second-order differential equation. However there is a short-cut which is worth using. We note that G has no explicit dependence on the variable z:

$$\left.\frac{\partial G}{\partial z}\right|_{f,f'=\text{const}} = 0. \tag{10.71}$$

It is a standard result that this condition implies the existence of a "first integral":

$$G - f'\frac{\partial G}{\partial f'} = \text{constant}. \tag{10.72}$$

(For a Lagrangian system where L does not depend on time this equation states that the hamiltonian is constant—i.e., energy is conserved!)

When we substitute the G of this problem into equation (10.72), we find:

$$f\sqrt{1 + f'^2} - f'\frac{ff'}{\sqrt{1 + f'^2}} = C, \qquad (10.73)$$

where C is a constant. This simplifies to $f/C = \sqrt{1 + f'^2}$, or

$$\frac{df}{dz} = \sqrt{(f/C)^2 - 1}. \qquad (10.74)$$

We can convert this into an indefinite integral:

$$\int dz = \int \frac{df}{\sqrt{(f/C)^2 - 1}}, \qquad (10.75)$$

which we can solve by substituting $f/C = \cosh u$:

$$z - z_0 = \int \frac{C\sinh u\, du}{\sqrt{\cosh^2 u - 1}} = \int C\, du = Cu, \qquad (10.76)$$

where z_0 is the constant of integration. Hence we find

$$f(z) = C\cosh\frac{z - z_0}{C}. \qquad (10.77)$$

Now we can use the boundary conditions to find z_0 and C. Since the film is symmetric about $d/2$, $z_0 = d/2$. The other condition is

$$f(0) = f(d) = R = C\cosh\frac{d}{2C}, \qquad (10.78)$$

a transcendental equation for C which we can solve with graphical or numerical methods. Note that if R/d is too small there is no solution, and the soap film will separate into two sheets, one on each ring.

Solution 1.9. a) The potential energy of the bead is

$$V = -mga\cos\theta. \qquad (10.79)$$

The kinetic energy, which separates into a term due to the bead's motion along the wire and a term due to the rotation of the bead with the wire, is

$$T = \frac{1}{2}ma^2\dot{\theta}^2 + \frac{1}{2}m\omega^2(a\sin\theta)^2. \tag{10.80}$$

The Lagrangian is $L = T - V$. Using Lagrange's equation,

$$\frac{d}{dt}\left(\frac{\partial L}{\partial \dot{\theta}}\right) - \frac{\partial L}{\partial \theta} = 0, \tag{10.81}$$

we find that

$$a\ddot{\theta} + g\sin\theta - a\omega^2\cos\theta\sin\theta = 0. \tag{10.82}$$

At an equilibrium point $\ddot{\theta} = 0$, so $g = a\omega^2\cos\theta$, or $\omega^2 = g/a\cos\theta$. This equation has a solution for ω only if $\omega^2 \geq g/a$, so the critical angular velocity is

$$\omega_c = \sqrt{\frac{g}{a}}, \tag{10.83}$$

and the equilibrium angle is

$$\theta_0 = \cos^{-1}\left(\frac{g}{a\omega^2}\right). \tag{10.84}$$

b) If the mass makes small oscillations around the equilibrium point θ_0, then we can describe the motion in terms of a small parameter $\phi = \theta - \theta_0$. The equation of motion (10.82) becomes

$$a\ddot{\phi} + g\sin(\theta_0 + \phi) - a\omega^2\cos(\theta_0 + \phi)\sin(\theta_0 + \phi) = 0. \tag{10.85}$$

Using standard trigonometric identities, the small angle approximations $\sin\phi \approx \phi$ and $\cos\phi \approx 1$, and our solution for θ_0 (10.84), it is easy to show that

$$\ddot{\phi} + \omega^2\left(1 - \frac{g^2}{a^2\omega^4}\right)\phi = 0. \tag{10.86}$$

This has the general solution

$$\phi = A\cos\Omega t + B\sin\Omega t, \tag{10.87}$$

where

$$\Omega = \omega\sqrt{1 - \frac{g^2}{a^2\omega^4}}, \qquad (10.88)$$

and A and B are arbitrary constants. The period of oscillation is $2\pi/\Omega$.

Solution 1.10. In plane-polar coordinates, the Lagrangian for a particle moving in a central potential $V(r)$ is

$$L = \frac{1}{2}m(\dot{r}^2 + r^2\dot{\theta}^2) - V(r), \qquad (10.89)$$

where m is the mass of the particle. The potential is given in the question as

$$V(r) = -\frac{k}{r} + \frac{1}{2}br^2. \qquad (10.90)$$

The θ-component of Lagrange's equation is

$$\frac{\partial L}{\partial \dot{\theta}} = mr^2\dot{\theta} = \text{constant} \equiv l. \qquad (10.91)$$

The hamiltonian of our system is then

$$H = \frac{p_r^2}{2m} + \frac{l^2}{2mr^2} + V(r) = \frac{p_r^2}{2m} + V_{\text{eff}}(r), \qquad (10.92)$$

with $p_r = m\dot{r}$ and

$$V_{\text{eff}}(r) = \frac{l^2}{2mr^2} + V(r). \qquad (10.93)$$

The term $l^2/2mr^2$ is referred to as an "angular momentum barrier." Solving the equations of motion for this hamiltonian is equivalent to solving Lagrange's equations for the Lagrangian:

$$L = \frac{1}{2}m\dot{r}^2 - V_{\text{eff}}(r). \qquad (10.94)$$

This is a completely general result for the motion of a particle in a central potential and could easily have been our starting point in this problem (e.g., Goldstein, Chapter 3).

It may seem unnecessarily long-winded to go through this procedure, but note that the sign of the angular momentum barrier in (10.94) is *opposite* to what we would have gotten if we had naively replaced $\dot{\theta}$ with l/mr^2 in the Lagrangian (10.89). This is due to the fact that the Lagrangian is a function of the time derivative of the position, and not of the canonical momentum.

The equation of motion from (10.94) is

$$m\ddot{r} = -\frac{d}{dr}V_{eff}(r). \tag{10.95}$$

If the particle is in a circular orbit at $r = r_0$ we require that the force on it at that radius should vanish,

$$\left.\frac{dV_{eff}}{dr}\right|_{r=r_0} = 0. \tag{10.96}$$

Using our expression for V_{eff} (10.93), we derive an expression relating the angular momentum l to the radius of the orbit r_0:

$$\frac{l^2}{mr_0^3} - \frac{k}{r_0^2} - br_0 = 0. \tag{10.97}$$

We are interested in perturbations about this circular orbit. Provided the perturbation remains small, we can expand $V_{eff}(r)$ about r_0,

$$V_{eff}(r) = V_{eff}(r_0) + (r - r_0)V'_{eff}(r_0) + \frac{1}{2}(r - r_0)^2 V''_{eff}(r_0) + \cdots . \tag{10.98}$$

If we use this expansion in the Lagrangian (10.94) together with the condition (10.96), we find

$$L = \frac{1}{2}m\dot{r}^2 - \frac{1}{2}(r - r_0)^2 V''_{eff}(r_0), \tag{10.99}$$

where we have dropped a constant term. This is just the Lagrangian for a simple harmonic oscillator, describing a particle undergoing radial oscillations with frequency

$$\omega^2 = \frac{1}{m}V''_{eff}(r_0). \tag{10.100}$$

Differentiating $V_{eff}(r)$ twice gives us

$$\frac{3l^2}{mr_0^4} - \frac{2k}{r_0^3} + b = m\omega^2. \tag{10.101}$$

We can eliminate l between equations (10.101) and (10.97) to give the frequency of radial oscillations:

$$\omega = \left(\frac{k}{mr_0^3} + \frac{4b}{m}\right)^{1/2}. \tag{10.102}$$

To find the rate of precession of the perihelion, we need to know the period of the orbit. From the definition of angular momentum l, equation (10.91), we have an equation for the orbital angular velocity ω_1,

$$\omega_1 \equiv \frac{d\theta}{dt} = \frac{l}{mr^2}. \tag{10.103}$$

Let us write $r(t) = r_0 + \epsilon(t)$, where $\epsilon(t)$ is sinusoidal with frequency ω and average value zero. We substitute $r(t)$ into equation (10.103) and expand in $\epsilon(t)$:

$$\frac{d\theta}{dt} = \frac{l}{mr_0^2}\left(1 - \frac{2\epsilon}{r_0} + \mathcal{O}(\epsilon^2)\right). \tag{10.104}$$

To zeroth order in the small quantities br_0^3/k and ϵ/r_0, the period of the orbit T_1 is the same as the period of oscillations $T_2 = 2\pi/\omega$. Therefore we can average ϵ over T_1 rather than T_2 and still get zero, to within terms of second order, which we are neglecting. The average angular velocity is therefore

$$\bar{\omega}_1 = \frac{2\pi}{T_1} \approx \frac{l}{mr_0^2} = \sqrt{\frac{k}{mr_0^3} + \frac{b}{m}}, \tag{10.105}$$

where we have made use of (10.97).

Now consider one complete period of the radial oscillation. This takes place in time $T_2 = 2\pi/\omega$. In this time the particle travels along its orbit through an angle of

$$\theta = 2\pi\frac{\bar{\omega}_1}{\omega} = 2\pi\frac{\sqrt{k/mr_0^3 + b/m}}{\sqrt{k/mr_0^3 + 4b/m}}$$

$$\approx 2\pi \left(1 - \frac{3br_0^3}{2k}\right). \qquad (10.106)$$

In other words, the particle does not quite orbit through 2π before the radial oscillation is completed. Each time around the perihelion precesses backwards through an angle

$$\delta\theta = 3\pi\frac{br_0^3}{k}, \qquad (10.107)$$

and it gets around in time T_2, so the precession rate is

$$\alpha = \frac{\delta\theta}{T_2} = \frac{3\pi br_0^3}{k}\frac{\sqrt{k/mr_0^3 + 4b/m}}{2\pi}$$

$$\approx \frac{3b}{2}\sqrt{\frac{r_0^3}{mk}}. \qquad (10.108)$$

b) When r is large enough that $F_r \approx -br$, we see that the force is like that of a linear spring. In this case the planar motion of the orbit can be resolved into simple harmonic motion in each of its three cartesian components. Thus the orbits will in general be ellipses; however, in each case the sun will be at the *center* of the ellipse rather than at one of the foci (as is the case for Newtonian gravity).

Chapter 11

Electricity & Magnetism—Solutions

Solution 2.1. This problem combines two of the simplest geometries which textbooks use to demonstrate the use of image charges to solve boundary value problems, so we should not be surprised that image charge methods work in this case as well. Let the line through the charge q and the center of the bulge be the $\hat{z}$-axis, with the origin such that the charge is at $z = p$ (Figure 11.1).

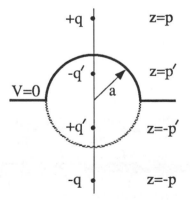

Figure 11.1.

We recall the rules of the image charge game: we want to replace the conductor with a distribution of "virtual charges" in such a way that the boundary conditions are satisfied. (See Jackson, Chapter 2.) If we ignore the plane for a moment and concentrate on the hemispherical bulge, it is easy to check that a second charge $-q'$, with $q' = qa/p$, at the point $z = p' = a^2/p$ leaves the surface of the bulge at a constant potential $V = 0$. This is the standard solution to the problem of a charge outside a spherical conductor. Now the trick is to introduce two more image charges: one of charge $-q$ at point $z = -p$, and one of charge q' at point $z = -p'$. These two charges do not change the potential on the hemispherical bulge, but now the potential also vanishes everwhere on the infinite plane.

The beauty of image-charge techniques now becomes apparant. The original charge feels the same force from a conducting infinite plane with a hemispherical bulge as it would from a set of three charges with the magnitudes and positions given above. So the force in the z-direction is given by

$$F = \sum_{q_i} \frac{qq_i}{r^2}, \tag{11.1}$$

where the sum is over the set of image charges. Our final answer is

$$
\begin{aligned}
F &= \frac{-qq'}{(p-p')^2} + \frac{qq'}{(p+p')^2} - \frac{q^2}{(2p)^2} \\
&= -q^2 \left(\frac{ap}{(p^2-a^2)^2} - \frac{ap}{(p^2+a^2)^2} + \frac{1}{4p^2} \right).
\end{aligned} \tag{11.2}
$$

Solution 2.2.   a) In order to make the assumption of noninteracting free charges more plausible, we can imagine there being a background of fixed charges $q = +e$, so that the plasma is neutral overall. Such a background could be provided by ions with mass $m_i \gg m$; then their contribution to the conductivity would be negligible.

We can write the incident plane wave in the form $\mathbf{E} = \mathbf{E}_0 e^{i(kx-\omega t)}$. The equation of motion for a single charge at position x is

$$m\ddot{\mathbf{x}} = -e\mathbf{E}, \tag{11.3}$$

which we integrate once to get

$$\dot{\mathbf{x}} = \frac{e\mathbf{E}}{im\omega}. \tag{11.4}$$

The current density carried by the charges is $\mathbf{j} = -ne\dot{\mathbf{x}}$. The conductivity can be found from Ohm's law, $\mathbf{j} = \sigma\mathbf{E}$:

$$\sigma = \frac{-ne^2}{im\omega}. \tag{11.5}$$

b) For sources in a vacuum, Maxwell's equations take the form:

$$\nabla \cdot \mathbf{E} = -4\pi ne, \tag{11.6}$$
$$\nabla \cdot \mathbf{B} = 0, \tag{11.7}$$
$$\nabla \times \mathbf{E} = -\frac{1}{c}\frac{\partial \mathbf{B}}{\partial t}, \tag{11.8}$$
$$\nabla \times \mathbf{B} = \frac{1}{c}\frac{\partial \mathbf{E}}{\partial t} + \frac{4\pi \mathbf{j}}{c}. \tag{11.9}$$

Taking the curl of equation (11.8), and using the identity

$$\nabla \times (\nabla \times \mathbf{E}) = \nabla(\nabla \cdot \mathbf{E}) - \nabla^2\mathbf{E}, \tag{11.10}$$

along with the assumption of constant density n we find

$$-\nabla^2\mathbf{E} = -\frac{1}{c}\frac{\partial}{\partial t}(\nabla \times \mathbf{B}). \tag{11.11}$$

We may use (11.9) and Ohm's law to write

$$\nabla^2\mathbf{E} = \frac{1}{c^2}\frac{\partial^2\mathbf{E}}{\partial t^2} + \frac{4\pi\sigma}{c^2}\frac{\partial \mathbf{E}}{\partial t}. \tag{11.12}$$

When we substitute in $\mathbf{E} = \mathbf{E}_0 e^{i(kx-\omega t)}$, this yields

$$k^2 = \frac{\omega^2}{c^2} + i\frac{4\pi\sigma\omega}{c^2}. \tag{11.13}$$

Using the result (11.5) for the conductivity gives

$$k = \frac{1}{c}\sqrt{\omega^2 - \omega_p^2}. \tag{11.14}$$

c) From this last result, it is easy to find the index of refraction:

$$n \equiv \frac{ck}{\omega} = \sqrt{1 - \frac{\omega_p^2}{\omega^2}}. \tag{11.15}$$

If $\omega < \omega_p$ the index of refraction is purely imaginary, and the wave cannot propagate in the plasma.

d) In the presence of a magnetic field, the equation of motion is

$$m\ddot{\mathbf{x}} = -e\mathbf{E} - \frac{e}{c}\dot{\mathbf{x}} \times \mathbf{B}_0, \tag{11.16}$$

where in this problem $\mathbf{B}_0 = B_0\hat{\mathbf{z}}$. We will investigate the propagation of left- and right-circularly polarized waves (LHCP and RHCP, respectively), traveling in the z-direction. For these two cases the electric field is

$$\begin{aligned} \mathbf{E} &= E_0 e^{i(kz-\omega t)}\left[\frac{1}{\sqrt{2}}(\hat{\mathbf{x}} \pm i\hat{\mathbf{y}})\right] \\ &\equiv E_0 e^{i(kz-\omega t)}\hat{\epsilon}_{\pm}, \end{aligned} \tag{11.17}$$

where $\hat{\epsilon}_+$ ($\hat{\epsilon}_-$) is the polarization vector for LHCP (RHCP) waves. We will use the basis $(\hat{\epsilon}_+, \hat{\epsilon}_-, \hat{\mathbf{z}})$, which has the useful relation $\hat{\epsilon}_{\pm} \times \hat{\mathbf{z}} = \pm i\hat{\epsilon}_{\pm}$.

There are no forces on the charges in the z-direction, so we can ignore that component in what follows. The position of the charge is given by $\mathbf{x} = x^+\hat{\epsilon}_+ + x^-\hat{\epsilon}_-$. Assuming the time dependence of $\mathbf{x}$ is $e^{i\omega t}$, and substituting (11.17) into the equation of motion (11.16), we find

$$-m\omega^2(x^+\hat{\epsilon}_+ + x^-\hat{\epsilon}_-) = -eE_0\hat{\epsilon}_{\pm}e^{-i\omega t} - \frac{\omega e B_0}{c}(x^+\hat{\epsilon}_+ - x^-\hat{\epsilon}_-), \tag{11.18}$$

where the upper (lower) sign applies for LHCP (RHCP) incident waves. This equation must be true for each component, so

$$x^{\pm} = \frac{eE_0}{m\omega(\omega \pm \omega_c)}e^{-i\omega t}, \tag{11.19}$$

provided that $\omega \neq \omega_c$, where $\omega_c = eB_0/mc$ is the cyclotron frequency. Taking the time derivative of $x^{\pm}$, and using $\mathbf{j} = -ne\dot{\mathbf{x}}$, we find

$$\sigma^{\pm} = \frac{-ne^2 E_0}{im(\omega \pm \omega_c)}.$$ (11.20)

Using this conductivity in the general dispersion relation (11.13) yields

$$n = \sqrt{1 - \frac{\omega_p^2}{\omega(\omega \pm \omega_c)}}.$$ (11.21)

Note that we chose to treat the plasma as consisting of noninteracting free charges in a vacuum, so that $\epsilon = 0$ and $\sigma \neq 0$. The same result for the index of refraction can be derived if we treat the plasma as a dielectric, with a complex $\epsilon \neq 0$ and a conductivity $\sigma = 0$.

The difference between the indices of refraction of left- and right-handed circularly polarized light leads to Faraday rotation, an effect used to measure the density of electrons in a magnetic field.

Solution 2.3. a) For an object of length L, cross-sectional area A, and conductance σ, the resistance is

$$R = \frac{L}{\sigma A} = \frac{L}{\sigma \pi b^2}.$$ (11.22)

This result is easily derived by comparing Ohm's law in the two forms $V = IR$ and $J = \sigma E$. Thus, the resistance of the cylinder is $L/\sigma_1 \pi b^2$.

b) We will treat the physical resistor with the defect as a group of ideal resistors. We make the following simplifying assumptions about the geometry of the defect:

- it is a cylinder of radius a and length a,

- concentric with the cylinder of the physical resistor, and

- and centered lengthwise on the center of the physical resistor.

Assume the defect can be viewed as an ideal resistor in parallel with the portion of the physical resistor which surrounds it. (See Figure 11.2.) We may then write

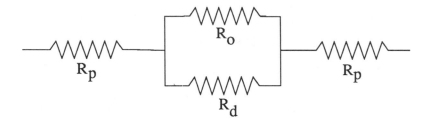

Figure 11.2.

$$R_{tot} = R_p + \left(\frac{1}{R_d} + \frac{1}{R_o}\right)^{-1} + R_p, \qquad (11.23)$$

where R_d is the resistance of the defect, R_p is the resistance of each of the two pieces of cylinder of length $(L/2 - a/2)$ that are "unaffected" by the presence of the defect, and R_o is the resistance of the cylinder of radius b missing a core cylinder of radius a. Using our formula for the resistance (11.22) we may immediately write

$$R_p = \frac{L/2 - a/2}{\sigma_1 \pi b^2}, \quad R_d = \frac{a}{\sigma_2 \pi a^2}, \quad \text{and} \quad R_o = \frac{a}{\sigma_1 \pi (b^2 - a^2)}. \qquad (11.24)$$

If we assume $b \gg a$ and crunch through the algebra, we find

$$R_{tot} \approx \frac{L}{\sigma_1 \pi b^2} - \frac{a^3(\sigma_2 - \sigma_1)}{\pi(\sigma_1 b^2)^2}, \qquad (11.25)$$

and the change in the resistance to first order is evidently

$$\delta R \approx \frac{a^3(\sigma_2 - \sigma_1)}{\pi(\sigma_1 b^2)^2}. \qquad (11.26)$$

c) To calculate the current in the defect, now considered spherical, it is simplest to solve for the potential everywhere. We set up a system of spherical coordinates, with the center of the defect at the origin, and measure θ from the axis of the resistor. The azimuthal symmetry allow us to expand the potential in the form:

$$\Phi(r, \theta, \phi) = \Phi(r, \theta) = \sum_l (A_l r^l + B_l r^{-(l+1)}) P_l(\cos \theta), \qquad (11.27)$$

where the $P_l(\cos \theta)$ are Legendre polynomials. (See Jackson, Chapter 3.) Let $\mathbf{E}_2$ be the electric field inside the defect and $\mathbf{E}_1$ the field outside. We have the following boundary conditions:

1. Within the main portion of the resistor, very far from the defect, the current distribution j_0 is uniform across the circular cross-section and is parallel to the axis of the resistor, the $\hat{\mathbf{z}}$-axis. Thus, $\mathbf{j}_1 \to j_0 \hat{\mathbf{z}}$ as $z \to \pm\infty$. This means that the electric field must satisfy the condition $\mathbf{E}_1 \to j_0 \hat{\mathbf{z}}/\sigma_1$ as $z \to \pm\infty$ (from the relation $\mathbf{j} = \sigma \mathbf{E}$), which gives us the following condition on the potential outside the defect:

$$\Phi_1 \longrightarrow -\frac{1}{\sigma_1} j_0 r \cos\theta \quad \text{as} \quad z \to \pm\infty. \qquad (11.28)$$

2. The electric field $\mathbf{E}_2$ inside the defect must be finite at the origin.

3. The tangential component of the electric field must be continuous across the boundary at $r = a$, because the curl of $\mathbf{E}$ must vanish.

4. We may use the continuity equation $\nabla \cdot \mathbf{j} = -\partial\rho/\partial t$ to see that $\nabla \cdot \mathbf{j} = 0$, since we have a steady current. Therefore, the normal component of $\mathbf{j}$ is continuous across the boundary, resulting in a discontinuity in the normal component of the electric field,

$$\sigma_1 E_{1_\perp} = \sigma_2 E_{2_\perp} \quad \text{at } r = a. \qquad (11.29)$$

From this we see that there is a charge build-up at the surface of the defect.

The first boundary condition leads to a big simplification. Because $\mathbf{j}$ is proportional to $\cos\theta$ as $z \to \pm\infty$, and $P_1(\cos\theta) = \cos\theta$, only the $l \le 1$ terms in the expansion of Φ will survive. Thus, we use the symmetry of the problem to set $B_l = 0$ for $l > 1$. Using (11.27), the most general potential commensurate with the first boundary condition is:

$$\Phi_1(r, \theta) = -\frac{j_0}{\sigma_1}\left(r + \frac{C}{r^2}\right)\cos\theta, \qquad (11.30)$$

$$\Phi_2(r, \theta) = \left(Dr + \frac{F}{r^2}\right)\cos\theta. \qquad (11.31)$$

The second boundary condition implies that Φ_2 may only have positive powers of r, so that $F = 0$. If we note that the tangential component of $\mathbf{E}$ is $E_\theta = -(1/r)(\partial\Phi/\partial\theta)$, the third boundary condition allows us to write

$$E_{2_\theta} = D\sin\theta = E_{1_\theta} = -\frac{j_0}{\sigma_1}(1 + \frac{C}{a^3})\sin\theta. \qquad (11.32)$$

The final boundary condition, where $E_\perp = E_r = -\partial\Phi/\partial r$, yields

$$\sigma_2 E_{2_\perp} = -\sigma_2 D\cos\theta = \sigma_1 E_{1_\perp} = j_0\left(1 - \frac{2C}{a^3}\right)\cos\theta. \qquad (11.33)$$

Now we use the last two equations to solve for D:

$$D = -\frac{3j_0}{\sigma_2 + 2\sigma_1}. \qquad (11.34)$$

Thus,

$$\Phi_2(r, \theta) = -\left(\frac{3j_0}{\sigma_2 + 2\sigma_1}\right)r\cos\theta. \qquad (11.35)$$

Finally, we write $\mathbf{j}_2 = \sigma_2\mathbf{E}_2 = -\sigma_2\nabla\Phi_2$ and find the current inside the defect in the resistor:

$$\mathbf{j}_2 = \left(\frac{3j_0}{1 + 2\sigma_1/\sigma_2}\right)(\hat{\mathbf{r}}\cos\theta - \hat{\theta}\sin\theta). \qquad (11.36)$$

Solution 2.4. a) In this simplified model of an antenna, the charge is zero everywhere except at the ends of the wire, $z = \pm l/2$, where there will be time-dependent charges $\pm Q(t)$, respectively. We can find $Q(t)$ from the current in the wire

$$\frac{dQ(t)}{dt} = I(t) = I_0 \cos \omega_0 t \,, \tag{11.37}$$

which we integrate to give

$$Q(t) = \frac{I_0}{\omega_0} \sin \omega_0 t \,. \tag{11.38}$$

The dipole moment $\mathbf{p}$ is $Q(t)$ times the separation of the charges,

$$\mathbf{p} = 2Q(t)\frac{l}{2}\hat{\mathbf{z}} = \frac{I_0 l}{\omega_0} \sin \omega_0 t \, \hat{\mathbf{z}} \,. \tag{11.39}$$

(Note that in a more realistic antenna problem, the current would not be a constant along the wire, and the charge density would be nonzero everywhere along the antenna.)

b) Let us choose the Lorentz gauge, in which the potentials satisfy

$$\nabla \cdot \mathbf{A} + \frac{1}{c}\frac{\partial \phi}{\partial t} = 0 \,. \tag{11.40}$$

The current density $\mathbf{J}$ in the antenna is

$$\mathbf{J}(\mathbf{x}) = \begin{cases} I_0 \cos \omega_0 t \delta(x)\delta(y)\hat{\mathbf{z}} & -l/2 < z < l/2, \\ 0 & \text{otherwise}, \end{cases} \tag{11.41}$$

and the vector potential $\mathbf{A}$ is

$$\mathbf{A}(\mathbf{x}) = \frac{1}{c}\int \mathbf{J}(\mathbf{x}')\frac{e^{ik|\mathbf{x}-\mathbf{x}'|}}{|\mathbf{x} - \mathbf{x}'|}d^3x'. \tag{11.42}$$

If $r \gg l$, we may approximate $|\mathbf{x} - \mathbf{x}'| \approx r - \hat{\mathbf{n}} \cdot \mathbf{x}'$, where $\hat{\mathbf{n}}$ is the unit vector along $\mathbf{x}$ and $|\mathbf{x}| = r$. We substitute this approximation for

$|\mathbf{x} - \mathbf{x}'|$ into the exponent. For the denominator it is sufficient to let $|\mathbf{x} - \mathbf{x}'| = r$. We then obtain

$$\mathbf{A}(\mathbf{x}) \approx \frac{e^{ikr}}{cr} \int \mathbf{J}(\mathbf{x}')e^{-ik\hat{\mathbf{n}}\cdot\mathbf{x}'}d^3x'. \tag{11.43}$$

Integrating over x' and y' and letting $\hat{\mathbf{n}} \cdot \hat{\mathbf{z}} = \cos\theta$ where θ is the angle between $\hat{\mathbf{n}}$ and the $\hat{\mathbf{z}}$-axis, we find

$$\mathbf{A}(\mathbf{x}) \approx \hat{\mathbf{z}}\frac{e^{ikr}}{cr} \int_{-l/2}^{l/2} I_0 e^{-ikz'\cos\theta} \cos\omega_0 t\, dz' \tag{11.44}$$

(where $\omega_0 = ck$), which integrates to give:

$$\mathbf{A}(\mathbf{x}) \approx \hat{\mathbf{z}}\frac{2I_0 e^{ikr}}{\omega_0 r \cos\theta} \sin\left(\frac{kl}{2}\cos\theta\right) \cos\omega_0 t. \tag{11.45}$$

(Note that for small kl, this reduces to the ordinary dipole far-field approximation: $\mathbf{A}(\mathbf{x}) \approx [I_0 l e^{ikr} \cos\omega_0 t]/cr$.)

Next we find the scalar potential ϕ, which is

$$\phi(\mathbf{x}) = \int \rho(\mathbf{x}')\frac{e^{ik|\mathbf{x}-\mathbf{x}'|}}{|\mathbf{x} - \mathbf{x}'|}d^3x'. \tag{11.46}$$

As in part (a), the charge density is zero everywhere except at the ends of the wire, where its integral is $\pm Q(t)$. Making the same approximations for $|\mathbf{x} - \mathbf{x}'|$ as above, we find

$$\phi(\mathbf{x}) \approx -\frac{2iI_0}{\omega_0 r}e^{ikr} \sin\left(\frac{kl}{2}\cos\theta\right) \sin\omega_0 t. \tag{11.47}$$

c) To obtain the radiation pattern, we need to examine the power per unit solid angle $dP/d\Omega$, which is proportional to $|\mathbf{B}|^2$. For $r \gg \lambda_0$, we note

$$\mathbf{B} = \nabla \times \mathbf{A} \approx ik\hat{\mathbf{n}} \times \mathbf{A}. \tag{11.48}$$

In spherical coordinates,

$$\hat{\mathbf{n}} = \hat{\mathbf{z}}\cos\theta + \hat{\mathbf{y}}\sin\theta\sin\phi + \hat{\mathbf{x}}\sin\theta\cos\phi, \tag{11.49}$$

which leads to

$$|\mathbf{B}|^2 \sim k^2 |A_z|^2 \sin^2 \theta. \tag{11.50}$$

Therefore, the radiation pattern is given qualitatively by

$$\frac{dP}{d\Omega} \propto \sin^2 \left(\frac{kl}{2} \cos \theta \right) \tan^2 \theta . \tag{11.51}$$

This may be contrasted with the ordinary dipole pattern, which is proportional to $\sin^2 \theta$.

Solution 2.5. We wish to find the angular velocity of the wheel $\omega(t)$ and the current in the circuit $I(t)$. First we define the current to be positive if it flows out of the positive terminal of the battery. Let the origin of our coordinate system be at the hub of the wheel and $\hat{z}$ be parallel to $\mathbf{B}$. Then if $\mathbf{r}$ gives the position along the current-carrying spoke of the wheel, the force on an infinitesimal element of this arm is

$$d\mathbf{F} = \frac{I}{c} d\mathbf{r} \times \mathbf{B}. \tag{11.52}$$

The torque on the wheel is therefore

$$J\dot\omega\,\hat{z} = \int \mathbf{r} \times d\mathbf{F} = \int_0^R \frac{IB}{c} r\,dr\,\hat{z}, \tag{11.53}$$

which leads to

$$J\dot\omega\hat{z} = \frac{IBR^2}{2c}\hat{z}. \tag{11.54}$$

We now have one equation relating $\omega(t)$ and $I(t)$. To find a second equation, we set the power delivered by the battery equal to the power absorbed by the rest of the circuit:

$$IV = I\left(L\dot{I}\right) + \frac{d}{dt}\left(\frac{1}{2}J\omega^2\right), \tag{11.55}$$

or

$$I\left(V - L\dot{I}\right) = J\omega\dot\omega. \tag{11.56}$$

Differentiating the torque equation (11.54) gives

$$\dot{I} = J\ddot{\omega}\left(\frac{2c}{BR^2}\right).$$

(11.57)

Substituting this expression into equation (11.56) and using (11.54) to cancel a common factor of I leads to a second order differential equation for $\omega(t)$:

$$\ddot{\omega} + \left(\frac{R^2 B}{2c}\right)^2 \frac{1}{LJ}\omega = \frac{R^2 B}{2cLJ}V.$$

(11.58)

This equation has the general solution

$$\omega(t) = C\cos\Omega t + D\sin\Omega t + \frac{2cV}{BR^2},$$

(11.59)

where C and D are constants to be determined from the initial conditions and

$$\Omega = \frac{R^2 B}{2c\sqrt{LJ}}.$$

(11.60)

To find the coefficient D we note that at $t = 0$ there is no current. Because $I(0) = 0$ and because $\dot{\omega}$ is proportional to $I(t)$ from equation (11.54), we have $\dot{\omega}(0) = 0$ and thus $D = 0$. To find C we note that at $t = 0$ the wheel is at rest, and thus $\omega(0) = 0$. Therefore $C = -2cV/BR^2$, and the final solution for $\omega(t)$ is

$$\omega(t) = \frac{2cV}{BR^2}(1 - \cos\Omega t).$$

(11.61)

We then use equation (11.54) to find

$$I(t) = \frac{V}{\Omega L}\sin\Omega t.$$

(11.62)

This problem can also be solved by the application of Kirchoff's law instead of the conservation of energy equation (11.55). When setting the sum of the voltage changes around the circuit equal to zero (Kirchoff's law), one must include the voltage induced by the changing flux through the circuit.

Solution 2.6. Since there are no currents in this problem, we may consider the cylinder to have a magnetic surface charge density on each end. It can be shown (see, for example, Jackson, Chapter 5) that the effective magnetic surface charge density σ is

$$\sigma = \hat{n} \cdot M, \qquad (11.63)$$

where $\hat{n}$ is the unit vector directed out of the surface. Since $L \gg R$, we may ignore the effective magnetic surface charge density on the end of the cylinder farthest from the infinitely permeable ceiling. The magnetic surface charge density σ at the end nearest the ceiling will induce an image surface charge density $-\sigma$. Let us displace the cylinder a distance ϵ from the infinitely permeable ceiling (Figure 11.3). The geometry is that of a parallel plate capacitor with a separation 2ϵ between the magnet's surface and the image charge, which appears a distance ϵ into the ceiling.

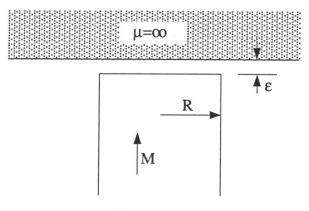

Figure 11.3.

In analogy with a parallel plate capacitor, the electric field caused by the image charge, evaluated at the position of the magnet's surface, is $E = 4\pi\sigma$. Thus the upward force on the magnetic surface charge density is $4\pi^2 R^2 \sigma^2 = 4\pi^2 R^2 M^2$.

We wish to find L such that this magnetic force just balances the gravitational force:

$$mg = \pi R^2 L \rho g = 4\pi^2 R^2 M^2, \qquad (11.64)$$

which leads to

$$L = 4\pi \frac{M^2}{g\rho}. \tag{11.65}$$

Solution 2.7. We have the circuit as shown in Figure 11.4. We need

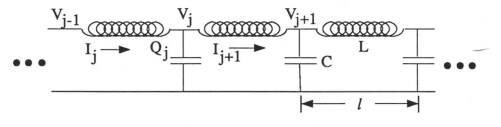

Figure 11.4.

the differential equation giving the behavior of $V(t, z = jl) \equiv V_j(t)$.
First we note that $\dot{Q}_j = C\dot{V}_j$, and $\dot{Q}_j + I_{j+1} = I_j$, so that

$$\dot{V}_j = \frac{1}{C}(I_j - I_{j+1}). \tag{11.66}$$

Furthermore, we can use the definition of inductance to write

$$\dot{I}_j L = V_{j-1} - V_j. \tag{11.67}$$

Taking the derivative of (11.66) and using equation (11.67) to eliminate
the current, we have

$$\ddot{V}_j = \frac{1}{LC}(V_{j-1} - 2V_j + V_{j+1}). \tag{11.68}$$

Now we assume that $V_j(t)$ is a periodic wave, or

$$V_j(t) = V_0 e^{i(jlk - \omega t)}, \tag{11.69}$$

where $k = 2\pi/\lambda$. Equation (11.68) then reduces to the desired dispersion relation:

$$\cos kl = \left(1 - \frac{\omega^2 LC}{2}\right). \tag{11.70}$$

The cutoff frequency is given by the condition

$$\left| 1 - \frac{\omega^2 LC}{2} \right| \leq 1, \tag{11.71}$$

which gives

$$\omega_c = \frac{2}{\sqrt{LC}}. \tag{11.72}$$

Waves with frequencies greater than ω_c will not be transmitted by the transmission line because k will have an imaginary component that will attenuate the wave.

Solution 2.8. A black stripe on a negative means that there was a maximum in the light intensity at that position. We want to find the positions of those maxima along the photographic plate.

a) The maxima will occur when the incident and reflected plane waves interfere constructively. Since there is a 180° phase shift upon reflection from the mirror, the condition for constructive interference is

$$\frac{2h}{\lambda}(2\pi) + \pi = n(2\pi), \tag{11.73}$$

where h is the height of a point of maximum intensity on the plate above the mirror and n is a positive integer. Solving for h gives

$$h = \frac{(2n-1)}{4}\lambda. \tag{11.74}$$

From Figure 11.5, $h = d\sin\alpha$, so

$$d = \frac{(2n-1)}{4}\frac{\lambda}{\sin\alpha}. \tag{11.75}$$

Thus, the black stripes are regularly spaced a distance $\lambda/(2\sin\alpha)$ apart, and the first is at a distance $\lambda/(4\sin\alpha)$ from the point of contact of the plate.

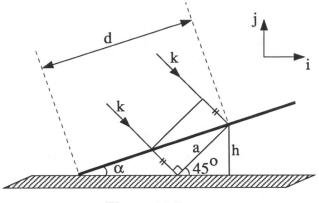

Figure 11.5.

b) The plane of incidence is defined by the k-vector of the incoming plane wave and the vector normal to the mirror's surface (Figure 11.5). Consider light polarized with $\mathbf{E}$ perpendicular to this plane. It will undergo a 180° phase shift upon reflection. The path difference between incident and reflected waves is $a = \sqrt{2}\, h$, so

$$\frac{\sqrt{2}h}{\lambda}(2\pi) + \pi = n(2\pi).\qquad(11.76)$$

Using $h = d \sin \alpha$, we find the positions of the black stripes:

$$d = \frac{2n-1}{2\sqrt{2}\sin\alpha}\lambda.\qquad(11.77)$$

The next case is trickier. If the light is polarized in the plane of incidence, $\mathbf{E}$ has components both parallel and perpendicular to the plane of the mirror, so one component undergoes a 180° phase shift, while the other does not. The incident $\mathbf{E}$ is at a 45° angle to the mirror, so we describe the time-independent portion of the incident wave as

$$\mathbf{E}_i = \frac{E_0}{\sqrt{2}}(\hat{\imath} + \hat{\jmath}),\qquad(11.78)$$

and the reflected piece as

$$\mathbf{E}_r = \frac{E_0}{\sqrt{2}}\left[\hat{\imath}\cos\left(\frac{2\pi a}{\lambda} + \pi\right) + \hat{\jmath}\cos\left(\frac{2\pi a}{\lambda}\right)\right],\qquad(11.79)$$

where $\hat{\imath}$ and $\hat{\jmath}$ are unit vectors in the plane of incidence, tangent and perpendicular to the mirror's surface, respectively. Adding these contributions together we find

$$\mathbf{E} = \frac{E_0}{\sqrt{2}} \left\{ \hat{\imath} \left[1 + \cos \left(\frac{2\pi a}{\lambda} + \pi \right) \right] + \hat{\jmath} \left[1 + \cos \left(\frac{2\pi a}{\lambda} \right) \right] \right\}. \qquad (11.80)$$

We are interested in the intensity, which is proportional to $|\mathbf{E}|^2$. We find

$$|\mathbf{E}|^2 = E_0^2 \left[1 + \cos^2 \left(\frac{2\pi a}{\lambda} \right) \right]. \qquad (11.81)$$

It is evident from the equation above that the intensity has a maximum when $2\pi a / \lambda = n\pi$, or $a = n\lambda/2$. Solving for d we find that the stripes are at positions given by

$$d = \frac{n\lambda}{2\sqrt{2}\sin \alpha}. \qquad (11.82)$$

Notice that all the black stripes present when the light was polarized perpendicular to the plane of incidence appear, as well as new stripes midway between the old ones.

———————

Solution 2.9. a) Since both the interior and exterior regions are free of sources, the electrostatic potential will satisfy Laplace's equation,

$$\nabla^2 \Phi = 0 \qquad (11.83)$$

everywhere except at $r = b$. The specification of the potential on the split cylinder and at infinity is sufficient to ensure the uniqueness of the solution.

The geometry of the problem dictates that we work in cylindrical coordinates (r, ϕ, z), with the split at $\phi = 0$ and at $\phi = \pi$. Laplace's equation (11.83) then has the form

$$\nabla^2 \Phi = \frac{1}{r} \frac{\partial}{\partial r} \left(r \frac{\partial \Phi}{\partial r} \right) + \frac{1}{r^2} \frac{\partial^2 \Phi}{\partial \phi^2} + \frac{\partial^2 \Phi}{\partial z^2} = 0. \qquad (11.84)$$

Our first step is to note that (by translational symmetry) our solution will be independent of z. In this case, the general solution of equation (11.84) is (see, for example, Jackson, Chapter 2)

$$\Phi(r, \phi) = a_0 + b_0 \ln r + \sum_{n=1}^{\infty} a_n r^n \sin(n\phi + \alpha_n) + \sum_{n=1}^{\infty} b_n r^{-n} \sin(n\phi + \beta_n).$$

$$(11.85)$$

Consider the interior region with potential $\Phi_{in}(r, \phi)$. We demand that the potential be well-defined at $r = 0$. This immediately tells us that $b_n = 0$ for all n including $n = 0$. A further simplification occurs if we note that with our choice of coordinates, the problem is antisymmetric under $\phi \to -\phi$. Hence we require that $a_0 = 0$ and $\alpha_n = 0$ for all n. Then the potential inside the cylinder takes the simpler form

$$\Phi_{in}(r, \phi) = \sum_{n=1}^{\infty} a_n r^n \sin n\phi. \qquad (11.86)$$

The boundary conditions are

$$\Phi(b, \phi) = \begin{cases} V_0 & 0 < \phi < \pi \\ -V_0 & -\pi < \phi < 0 \end{cases}. \qquad (11.87)$$

To solve for Φ_{in} , we take the series expansion (11.86), set $r = b$, multiply both sides by $\sin m\phi$, and integrate over the range $-\pi < \phi \leq \pi$:

$$\int_{-\pi}^{\pi} d\phi \sin m\phi \, \Phi_{in}(b, \phi) = \sum_{n=1}^{\infty} a_n b^n \int_{-\pi}^{\pi} d\phi \sin n\phi \, \sin m\phi = \pi a_m b^m.$$

$$(11.88)$$

If we substitute the boundary conditions from (11.87) and do the integration, we get

$$a_m = \frac{-2V_0}{\pi b^m m} [(-1)^m - 1] = \begin{cases} 0, & m \text{ even} \\ 4V_0/\pi m b^m, & m \text{ odd} \end{cases}. \qquad (11.89)$$

This gives us the solution for the interior region:

$$\Phi_{in} = \frac{4V_0}{\pi} \sum_{n=0}^{\infty} \left(\frac{r}{b}\right)^{2n+1} \frac{\sin(2n+1)\phi}{2n+1}. \qquad (11.90)$$

The procedure to find the exterior potential is entirely analogous, except that we now demand that Φ_{ext} be well-behaved as $r \to \infty$. This implies that $b_0 = 0$ and $a_n = 0$ for $n \geq 1$. The result is that

$$\Phi_{ext} = \frac{4V_0}{\pi} \sum_{n=0}^{\infty} \left(\frac{b}{r}\right)^{2n+1} \frac{\sin(2n+1)\phi}{2n+1}. \tag{11.91}$$

b) To find the surface charge density $\sigma(\phi)$ we use the standard result (easily derived from Gauss's law) that

$$(\mathbf{E}_2 - \mathbf{E}_1) \cdot \hat{\mathbf{n}} = 4\pi\sigma, \tag{11.92}$$

where $\mathbf{E}_1$ and $\mathbf{E}_2$ are the electric fields just inside and just outside the cylinder, and $\hat{\mathbf{n}} = \hat{\mathbf{r}}$ is the unit vector normal to the surface of the cylinder.

Since $\mathbf{E} = -\nabla\Phi$ and $\mathbf{E} \cdot \hat{\mathbf{n}} = -\left.\partial\Phi/\partial r\right|_{r=b}$, the left-hand side of equation (11.92) is simply the discontinuity in $\partial\Phi/\partial r$ across $r = b$, and the surface charge density is given by

$$\sigma = -\frac{1}{4\pi} \left[\frac{\partial\Phi_{ext}}{\partial r} - \frac{\partial\Phi_{in}}{\partial r}\right]_{r=b}. \tag{11.93}$$

If we substitute into this potentials (11.90) and (11.91), we find

$$\sigma = \frac{2\,V_0}{b\,\pi^2} \sum_{n=0}^{\infty} \sin(2n+1)\phi. \tag{11.94}$$

To sum this infinite series we can use a trick which we will later justify. First we write the sine as a sum of two exponentials and note that we have two geometric progressions:

$$\sum_{n=0}^{\infty} \sin(2n+1)\phi = \sum_{n=0}^{\infty} \frac{1}{2i}\left(e^{(2n+1)i\phi} - e^{-(2n+1)i\phi}\right). \tag{11.95}$$

As they stand, the sums are not well defined, as the individual terms do not go to zero. However, if we add a small imaginary constant $i\delta$

to ϕ in the first series and subtract it in the second, we can do the summation:

$$\sum_{n=0}^{\infty} \left(e^{(2n+1)(i\phi-\delta)} - e^{-(2n+1)(i\phi+\delta)} \right) = \left[\frac{e^{i\phi-\delta}}{1 - e^{2i\phi-2\delta}} - \frac{e^{-i\phi-\delta}}{1 - e^{-2i\phi-2\delta}} \right].$$

(11.96)

Physically, adding a small constant $i\delta$ to ϕ is equivalent to summing the series in equation (11.93) *before* taking the limit $r \to b$, in which case all the sums converge. We can now take the limit $\delta \to 0$ to evaluate the sum:

$$\sum_{n=0}^{\infty} \sin(2n+1)\phi = \frac{1}{2 \sin \phi}.$$

(11.97)

Therefore the charge density distribution is

$$\sigma = \frac{V_0}{\pi^2 b \sin \phi}.$$

(11.98)

c) So far we have ignored the finite separation ϵ of the plates. This separation becomes important if we integrate the charge distribution (11.98) to get the total charge per unit length on each plate; the integral is logarithmically divergent at $\phi = 0$ and π.

Since the plates are ϵ apart, and the cylinder has radius b, we let ϕ run in the range

$$\epsilon/2b \le \phi < \pi - \epsilon/2b$$

(11.99)

for the top plate, and

$$-\pi + \epsilon/2b < \phi \le -\epsilon/2b$$

(11.100)

for the bottom plate. The capacitance per unit length is $C \equiv Q/V$. Here V is the voltage difference, $2V_0$, and Q is the charge per unit length on each plate. We have

$$Q = \int_{\epsilon/2b}^{\pi-\epsilon/2b} b \, d\phi \, \sigma(\phi) = \frac{V_0}{\pi^2} \int_{\epsilon/2b}^{\pi-\epsilon/2b} \frac{d\phi}{\sin \phi} = \frac{2V_0}{\pi^2} \int_{\epsilon/2b}^{\pi/2} \frac{d\phi}{\sin \phi}. \quad (11.101)$$

We make the substitution $u = \cos \phi$ to find

$$\begin{aligned} Q &= \frac{2V_0}{\pi^2} \int_0^{u_0} \frac{du}{1-u^2} = \frac{2V_0}{\pi^2} \int_0^{u_0} du \, \frac{1}{2} \left[\frac{1}{1+u} + \frac{1}{1-u} \right] \\ &= \frac{V_0}{\pi^2} \ln \frac{1+u_0}{1-u_0}, \end{aligned}$$

(11.102)

where $u_0 = \cos \epsilon / 2b \approx 1 - \epsilon^2 / 8b^2$. We may then solve for the capacitance per unit length:

$$C = \frac{Q}{2V_0} \approx \frac{1}{2\pi^2} \ln \frac{1 + 1 - \epsilon^2/8b^2}{\epsilon^2/8b^2} \approx \frac{1}{2\pi^2} \ln \frac{16b^2}{\epsilon^2}, \qquad (11.103)$$

which may be simplified to give

$$C \approx \frac{1}{\pi^2} \ln \frac{4b}{\epsilon}. \qquad (11.104)$$

Solution 2.10. a) For nonrelativistic motion of an accelerating electron with velocity $v = \beta c$, the electric component of the radiation field is (see Jackson, Chapter 14)

$$\mathbf{E} = \frac{-e}{c} \left[\frac{\hat{\mathbf{n}} \times (\hat{\mathbf{n}} \times \dot{\boldsymbol{\beta}})}{R} \right]_{\text{ret}}, \qquad (11.105)$$

at a distance R in the direction $\hat{\mathbf{n}}$ from the electron. The expression in brackets is evaluated at the retarded time $t' = t - R/c$.

The magnetic radiation field is given by $\mathbf{B} = \hat{\mathbf{n}} \times \mathbf{E}$, and the energy flux by the Poynting vector:

$$\mathcal{S} = \frac{c}{4\pi} (\mathbf{E} \times \mathbf{B}) = \frac{c}{4\pi} |\mathbf{E}|^2 \hat{\mathbf{n}}. \qquad (11.106)$$

The angular dependence in $|\mathbf{E}|$ comes from the cross-product $\hat{\mathbf{n}} \times \dot{\boldsymbol{\beta}}$. Let θ be the angle between the line connecting the electron and the nucleus, and the line connecting the electron and the observation point (at time t'). Then $\hat{\mathbf{n}} \times \dot{\boldsymbol{\beta}}$ is proportional to $\sin \theta$, and from (11.106) the radiated power varies as $\sin^2 \theta$.

b) To find the polarization of the radiation, we observe that the vector $\hat{\mathbf{n}} \times \dot{\boldsymbol{\beta}}$ defines the normal to the plane containing the electron, the nucleus and the observer. From equation (11.105) we see that the electric field is perpendicular to this normal — i.e., it is in the plane

— and also that it is perpendicular to the line between the charge and the observer. This is sufficient to define its polarization. Alternatively, we could write the (unnormalized) polarization vector as $\epsilon = -\hat{n} \times (\hat{n} \times \hat{z})$, where $\hat{z}$ is a unit vector pointing from the electron to the proton.

c) For $v \ll c$ we can ignore the complication of having to evaluate the fields at retarded time. So

$$|\mathbf{E}| \approx \frac{e}{c} \frac{|\hat{n} \times (\hat{n} \times \dot{\beta}\,)|}{R} = \frac{e}{cR} |\dot{\beta} \sin \theta|, \qquad (11.107)$$

and therefore the Poynting vector is

$$S \approx \frac{e^2}{4\pi cR^2} \dot{\beta}^2 \sin^2 \theta \, \hat{n}. \qquad (11.108)$$

We can integrate this over a sphere of radius R centered on the charge to get the total power radiated:

$$
\begin{aligned}
\mathcal{P} &= \int d\Omega \, R^2 \, S \cdot \mathbf{n} = \int d\Omega \, \frac{e^2 a^2}{4\pi c^3} \sin^2 \theta \\
&= \frac{e^2 a^2}{4\pi c^3} \int_0^{2\pi} d\phi \int_{-1}^{1} d(\cos \theta) \, (1 - \cos^2 \theta) = \frac{2e^2 a^2}{3c^3} \quad (11.109)
\end{aligned}
$$

(where $a/c = \dot{\beta}$), which is the well-known Larmor formula.

By assumption we are neglecting the force on the electron due to radiation reaction, and so the acceleration of the electron is due simply to the electrostatic force between it and the nucleus, $F = Ze^2/r^2$. Hence our expression for the energy radiated per unit time when the electron is at a distance r from the nucleus is

$$P(r) = \frac{2e^2}{3c^3} \left(\frac{Ze^2}{mr^2}\right)^2 = \frac{2Z^2 e^6}{3m^2 c^3 r^4}, \qquad (11.110)$$

where m is the mass of the electron.

d) To evaluate the total power loss, we have to use the simplifying assumption that the total energy radiated is small enough that we may ignore its effect on the motion of the electron. (In fact this is no more than restating that we are ignoring radiation reaction.) Hence we can use conservation of energy,

$$-\frac{Ze^2}{r_0} = \frac{1}{2}mv^2 - \frac{Ze^2}{r}, \tag{11.111}$$

to find the velocity squared:

$$v^2 = \frac{2Ze^2}{m}\left(\frac{1}{r} - \frac{1}{r_0}\right). \tag{11.112}$$

The total energy radiated is

$$\mathcal{E} = \int dt \, \mathcal{P} = \int dr' \left|\frac{dt}{dr'}\right| \mathcal{P}(r'), \tag{11.113}$$

where $dt/dr = 1/v$, which is less than zero because the particle is falling inwards. Substituting for the speed of the particle from equation (11.112) gives us

$$\mathcal{E} = \sqrt{\frac{m}{2Ze^2}} \int_r^{r_0} \frac{dr'}{(1/r' - 1/r_0)^{1/2}} \frac{2Z^2 e^6}{3m^2 c^3 r'^4}. \tag{11.114}$$

We have to evaluate the integral

$$I = \int_r^{r_0} \frac{dr'}{(1/r' - 1/r_0)^{1/2}} \frac{1}{r'^4}. \tag{11.115}$$

If we make the substitution $1/r' - 1/r_0 = w^2$, $dr'/r'^2 = -2w \, dw$ and define $w_0^2 \equiv 1/r - 1/r_0$, then this becomes

$$I = \int_{w_0}^0 \frac{-2w \, dw}{w}\left(w^2 + \frac{1}{r_0}\right)^2 \tag{11.116}$$

$$= 2\left[\frac{1}{r_0^2}\left(\frac{1}{r} - \frac{1}{r_0}\right)^{1/2} + \frac{2}{3r_0}\left(\frac{1}{r} - \frac{1}{r_0}\right)^{3/2} + \frac{1}{5}\left(\frac{1}{r} - \frac{1}{r_0}\right)^{5/2}\right].$$

Thus the total radiated power is

$$\mathcal{E} = \left(\frac{2Z}{m}\right)^{3/2} \frac{e^5}{45c^3 r_0^{5/2}}\left(\frac{r_0}{r} - 1\right)^{1/2}\left[8 + \frac{4r_0}{r} + \frac{3r_0^2}{r^2}\right]. \tag{11.117}$$

Chapter 12

Quantum Mechanics—Solutions

Solution 3.1. a) We assume the reader is familiar with the separation of the wavefunction into angular and radial parts when the potential is spherically symmetric. The result is the radial equation for a modified radial wavefunction $u(r) = rR(r)$, where $\Psi(r, \theta, \phi) = f(\theta, \phi)R(r)$:

$$\frac{-\hbar^2}{2m}\frac{d^2}{dr^2}u(r) + \left[V(r) + \frac{l(l+1)\hbar^2}{2mr^2}\right]u(r) = Eu(r). \qquad (12.1)$$

Naturally, the condition for the minimum value of c will arise when there is only one bound state, which will obviously be the ground state, with $l = 0$. Since $V \to 0$ as $r \to \infty$, the bound states have $E < 0$, so we are left with (for $r \neq a$)

$$\frac{d^2}{dr^2}u(r) - k^2u(r) = 0, \qquad (12.2)$$

where $k^2 = 2m|E|/\hbar^2$.

Denote the regions $r < a$ and $r > a$ as regions I and II, respectively. The wavefunction in region I is

$$u_I(r) = A\sinh kr, \qquad (12.3)$$

because $u(0) = 0$, and in region II

$$u_{II}(r) = Be^{-kr}, \tag{12.4}$$

so that $u(r)$ vanishes as $r \to \infty$. Since we require the wavefunction to be continuous across the delta function at $r = a$,

$$A \sinh ka = Be^{-ka}. \tag{12.5}$$

The condition on the derivative of $u(r)$ is found by integrating the radial equation across the delta function:

$$\frac{-\hbar^2}{2m} \int_{a-\epsilon}^{a+\epsilon} \frac{d^2}{dr^2} u(r) dr - c \int_{a-\epsilon}^{a+\epsilon} \delta(r-a) u(r) dr = \int_{a-\epsilon}^{a+\epsilon} Eu(r) dr. \tag{12.6}$$

The integrations give

$$u'(a+\epsilon) - u'(a-\epsilon) = -\lambda u(a) - \mathcal{O}(\epsilon), \tag{12.7}$$

where $\lambda = 2mc/\hbar^2$. Letting $\epsilon \to 0$ gives

$$u'_{II}(a) - u'_I(a) = -\lambda u(a). \tag{12.8}$$

Taking the derivatives, using (12.5), and rearranging the terms leads to a transcendental equation,

$$\coth ka = \left(\frac{\lambda a}{ka} - 1\right). \tag{12.9}$$

The curves in Figure 12.1 must cross in order for there to be a solution. Consider the region $x \equiv ka \to 0$. For small x we can write

$$\coth x = \frac{1}{x} \left[1 + \mathcal{O}(x^2)\right], \tag{12.10}$$

so that, instead of the transcendental equation (12.9) we need to solve

$$\frac{1}{x} = \frac{\lambda a}{x} - 1, \tag{12.11}$$

or $x = \lambda a - 1$. In order for there to be a solution with $x > 0$, we require $\lambda a - 1 > 0$, or $c > \hbar^2/2ma$.

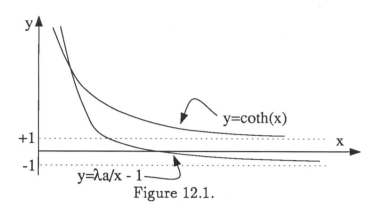

Figure 12.1.

b) In the limit of small incident velocity, we can consider a partial-wave analysis where only the $l = 0$ term is important. In this limit the scattering will be isotropic.

To find the cross-section, we will need the following results from the method of partial waves:

$$f(\theta) = \frac{1}{k} \sum_{l=0}^{\infty} (2l+1) e^{i\delta_l} \sin \delta_l \, P_l(\cos \theta), \qquad (12.12)$$

$$\sigma(\theta) = |f(\theta)|^2. \qquad (12.13)$$

Since we are ignoring all but the $l = 0$ term, we proceed to find the $l = 0$ phase shift. First we write down $u(r)$ in regions I and II,

$$u_I(r) = A \sin kr, \qquad (12.14)$$

$$u_{II}(r) = B (\sin kr + \tan \delta_0 \cos kr), \qquad (12.15)$$

where the form for $u_{II}(r)$ was chosen for its convenience and is equivalent to the form $\sin(kr + \delta_0)$. Once again we match the wavefunction and evaluate the discontinuity in the derivative at $r = a$ using equation (12.8), finding eventually

$$\tan \delta_0 = \frac{\lambda \sin^2 ka}{k - \lambda \cos ka \sin ka}. \qquad (12.16)$$

In the limit of small k, $\sin ka \approx ka$ and $\cos ka \approx 1$. Taking this limit we find

$$\tan \delta_0 \approx \frac{\lambda ka^2}{1 - \lambda a}. \qquad (12.17)$$

Using the expression

$$f_0 = \frac{1}{k} e^{i\delta_0} \sin \delta_0 = \frac{1}{k} \frac{\tan \delta_0}{1 - i \tan \delta_0},$$

(12.18)

and using the $l = 0$ term from the partial-wave cross-section, $\sigma = 4\pi |f_0|^2$, we find that

$$\sigma = 4\pi |f_0|^2 = \frac{4\pi \lambda^2 a^4}{\lambda^2 k^2 a^4 + (1 - \lambda a)^2}$$

(12.19)

Solution 3.2. a) In the space between the walls we have a free particle, so we may separate variables and write the wavefunction as a sum of plane waves,

$$\psi = \sum_{\mathbf{k}} A_{\mathbf{k}} e^{\pm i k_x x} e^{\pm i k_y y} e^{\pm i k_z z}.$$

(12.20)

This wavefunction is subject to the boundary conditions $\psi = 0$ at $x = 0$ and $x = D$. This leads to

$$\psi = \sum_{\mathbf{k}} A_{\mathbf{k}} \sin k_x x\, e^{\pm i k_y y} e^{\pm i k_z z},$$

(12.21)

where $k_x D = n\pi$. The ground state has $k_x = \pi/D$ and $k_y = k_z = 0$, and thus the energy of the ground state is

$$E_{gs} = \frac{\hbar^2 \pi^2}{2MD^2}.$$

(12.22)

b) An adiabatic change means a change such that the particle remains in an energy eigenstate with the same quantum numbers (here, the ground state). This means we may replace D with $2D$ in our expression for E_{gs} in order to find the subsequent energy:

$$E'_{gs} = \frac{\hbar^2 \pi^2}{8MD^2}.$$

(12.23)

The change in energy is therefore

$$\Delta E = \frac{3}{4}\left(\frac{\hbar^2\pi^2}{2MD^2}\right) = \frac{3}{4}E_{gs}. \tag{12.24}$$

Now we will do the classical calculation. Suppose that at some instant the walls are a distance s apart, and the ball is bouncing with speed v between the walls. For simplicity we will consider one wall fixed. If the other wall is moving slowly, the ball will hit it at a rate $2s/v$, and the average force on the wall will be

$$F = 2Mv\frac{v}{2s} = \frac{Mv^2}{s}. \tag{12.25}$$

The energy lost by the ball during the expansion is $\mathcal{E} = \int F\,ds$, and this is equal to the change in its kinetic energy, $\mathcal{E} = \frac{1}{2}M(v_0^2 - v^2)$. We differentiate to get $F = d\epsilon/ds$, so

$$\frac{Mv^2}{s} = \frac{d}{ds}\frac{1}{2}M(v_0^2 - v^2) = -Mv\frac{dv}{ds}, \tag{12.26}$$

or $dv/ds = -v/s$. We solve this to find $v(s) = v_0 D/s$. This shows us that when the walls are a distance $s = 2D$ apart, $v = v_0/2$ and the change in energy is

$$\Delta E = \frac{1}{2}M\left(v_0^2 - \frac{1}{4}v_0^2\right) = \frac{3}{4}E_0, \tag{12.27}$$

which is identical to the quantum-mechanical result.

c) In the case that one wall is moved rapidly, we can make use of the "sudden approximation" (see, for example, Messiah, Vol. II). At the moment that the change is made, the wavefunction will be unchanged for $x < D$, and zero for $x > D$ (See Figure 12.2):

$$\psi(x) = \begin{cases} \sqrt{\frac{2}{D}}\sin\frac{\pi x}{D} & 0 < x < D \\ 0 & D < x < 2D \end{cases}. \tag{12.28}$$

If we calculate the energy, i.e., if we take the expectation value of the operator $-\hbar^2\nabla^2/2M$, it is clear that for $x < D$ we get the same as in

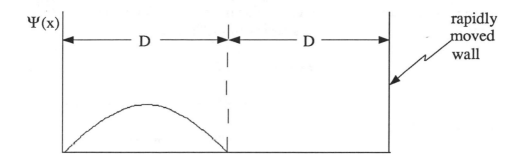

Figure 12.2.

part (a) and for $x > D$ we get zero. Hence the expectation value of the energy is unchanged.

However, the particle is no longer in an energy eigenstate, but in a superposition of eigenstates of a system with two walls that are separated by $2D$. The ground-state wavefunction for this separation is

$$\phi = \frac{1}{\sqrt{D}} \sin \frac{\pi x}{2D}, \qquad (12.29)$$

and the probability that the ball is in this state is given by the square of the overlap a where

$$a = \int_0^D dx \left(\sqrt{\frac{2}{D}} \sin \frac{\pi x}{D} \right) \left(\sqrt{\frac{1}{D}} \sin \frac{\pi x}{2D} \right) = \frac{4\sqrt{2}}{3\pi}. \qquad (12.30)$$

Therefore the probability of the particle being left in the ground state is $P = 32/9\pi^2$.

Solution 3.3. a) The general form of the Hamiltonian in the presence of an electromagnetic field is

$$H = \frac{1}{2m} \left(\mathbf{p} - \frac{e}{c}\mathbf{A} \right)^2 + e\phi, \qquad (12.31)$$

where $\mathbf{A}$ and ϕ are the vector and scalar potentials. A convenient choice of gauge which minimizes cross terms in (12.31) is

$$\mathbf{A} = (Bz, 0, 0), \quad \phi = -Ez, \tag{12.32}$$

where we can check that $\mathbf{B} = \nabla \times \mathbf{A}$ and $\mathbf{E} = -\nabla \phi$. The Schrödinger equation in this gauge is

$$H\psi = \left[\frac{1}{2m} \left\{ \left(p_x - \frac{e}{c}Bz \right)^2 + p_y^2 + p_z^2 \right\} - eEz \right] \psi = \mathcal{E}\psi, \tag{12.33}$$

with $\mathcal{E}$ the energy.

b) To separate variables, we note that equation (12.33) has no terms involving either x or y, which suggests a simple solution for these two variables. It is easy to check that the solutions in the x- and y-directions are plane waves, so we write

$$\psi(x, y, z) = e^{ik_x x + ik_y y} \phi(z). \tag{12.34}$$

Substituting this into the Schrödinger equation gives

$$\left[\frac{1}{2m} \left\{ \left(\hbar k_x - \frac{e}{c}Bz \right)^2 + \hbar^2 k_y^2 + p_z^2 \right\} - eEz \right] \phi = \mathcal{E}\phi, \tag{12.35}$$

which is a one-dimensional problem.

c) Rearranging terms and collecting the constants into $\mathcal{E}'$ gives us the equation

$$\left\{ p_z^2 + \left(\frac{eB}{c}z - \hbar k_x - \frac{mEc}{B} \right)^2 \right\} \phi = \mathcal{E}'\phi, \tag{12.36}$$

which we recognize to be that of a simple harmonic oscillator (centered around a point other than the origin). The expectation value of z in this case is simply the position z where the potential is a minimum:

$$\langle z \rangle = \frac{c}{eB} \left(\hbar k_x + \frac{mEc}{B} \right). \tag{12.37}$$

We want to find the expectation value of v_x. Using the standard result that $i\hbar v \equiv i\hbar \, dx/dt = [x, H]$ and the commutation relation $[x, p_x] = i\hbar$ we find

$$\langle v_x \rangle = \frac{1}{m}\left(\langle p_x \rangle - \frac{eB}{c}\langle z \rangle\right) = -\frac{Ec}{B}, \qquad (12.38)$$

where we have used $\langle p_x \rangle = \hbar k_x$.

We recognize this as the classical result, found by requiring that the total electromagnetic force on the particle in the x-direction vanish:

$$F_x = e\left(\mathbf{E} + \frac{\mathbf{v}}{c} \times \mathbf{B}\right) \cdot \hat{\mathbf{x}} = 0. \qquad (12.39)$$

Solution 3.4. We can write the Hamiltonian as $H = H_0 + V'$, where

$$
\begin{aligned}
H_0 &= \frac{\mathbf{p}^2}{2m} + \frac{1}{2}k(x^2 + y^2 + z^2), \text{ and} & (12.40) \\
V' &= qAe^{-(t/\tau)^2}z, & (12.41)
\end{aligned}
$$

and where $V'(t)$ is assumed to be small.

It is usually easier to solve problems involving a simple harmonic oscillator potential using raising and lowering operators. We can write the unperturbed hamiltonian H_0 as

$$H_0 = \hbar\omega(a_x^\dagger a_x + a_y^\dagger a_y + a_z^\dagger a_z + \frac{3}{2}), \qquad (12.42)$$

where we have defined:

$$
\begin{aligned}
\omega &\equiv \sqrt{\frac{k}{m}}, & (12.43) \\
a_x^\dagger &\equiv \left(\frac{m\omega}{2\hbar}\right)^{1/2}\left(x - \frac{i}{m\omega}p_x\right), & (12.44) \\
a_x &\equiv \left(\frac{m\omega}{2\hbar}\right)^{1/2}\left(x + \frac{i}{m\omega}p_x\right), & (12.45)
\end{aligned}
$$

with analogous definitions for a_y, $a_y^\dagger$, a_z, and $a_z^\dagger$. The operators a and $a^\dagger$ are the annihilation and creation operators from which we can form the number operator,

$$a^\dagger a \,|n\rangle = n\,|n\rangle, \qquad (12.46)$$

where n is some integer. The eigenstates of H_0 are therefore

$$|\mathbf{n}\rangle = |n_x, n_y, n_z\rangle, \qquad (12.47)$$

where n_x, n_y and n_z are integers. The energies are given by

$$H_0\,|n_x, n_y, n_z\rangle = \hbar\omega(n_x + n_y + n_z + \frac{3}{2})\,|n_x, n_y, n_z\rangle. \qquad (12.48)$$

Using the eigenstates of H_0 as a basis we can write an arbitrary wavefunction as

$$|\psi\rangle = \sum_{\mathbf{n}} c_{\mathbf{n}}(t)\,|\mathbf{n}\rangle e^{-iE_{\mathbf{n}}t/\hbar}, \qquad (12.49)$$

where the $c_{\mathbf{n}}(t)$ are complex coefficients. If the initial state at $t = -\infty$ is $|s\rangle$, then $c_s(-\infty) = 1$. According to time-dependent perturbation theory at $t = +\infty$, to first order in the perturbing potential V',

$$|c_{\mathbf{n}}(+\infty)|^2 = \frac{1}{\hbar^2}\left|\int_{-\infty}^{+\infty} V'_{\mathbf{ns}}(t')e^{i\omega_{\mathbf{ns}}t'}dt'\right|^2, \qquad (12.50)$$

where $V'_{\mathbf{ns}} = \langle\mathbf{n}\,|V'|\,\mathbf{s}\rangle$, and $\omega_{\mathbf{ns}} = (E_{\mathbf{n}} - E_{\mathbf{s}})/\hbar$. This result is not hard to derive from Schrödinger's equation if we write the wavefunction in the form (12.49). In this problem, $\mathbf{s}$ labels the ground state: $\mathbf{s} = (0,0,0)$. Therefore the probability that the system is in any excited state at $t = +\infty$ is given by the sum

$$P = \sum_{\mathbf{n}\neq\mathbf{s}} |c_{\mathbf{n}}(+\infty)|^2. \qquad (12.51)$$

To evaluate this sum, we need the matrix elements of $V'(t)$, which are

$$\langle n_x, n_y, n_z\,|V'(t)|\,0,0,0\rangle = qAe^{-(t/\tau)^2}\langle n_x, n_y, n_z\,|z|\,0,0,0\rangle. \qquad (12.52)$$

Rewriting z in terms of the raising and lowering operators, we can see that V' only connects states whose values of n_z differ by one, so that

$$V'_{\mathbf{ns}} = qAe^{-(t/\tau)^2}\left(\frac{\hbar}{2m\omega}\right)^{\frac{1}{2}}\delta_{n_z,1}\,\delta_{n_x,0}\,\delta_{n_y,0}, \qquad (12.53)$$

and only one term in the sum is nonzero. The desired probability is

$$P = \frac{1}{2\hbar m \omega} q^2 A^2 |I|^2, \tag{12.54}$$

with

$$I = \int_{-\infty}^{+\infty} e^{-(t/\tau)^2} e^{i\omega t} dt. \tag{12.55}$$

This integral can be evaluated by substituting $u = t/\tau$, completing the square in the exponent, and evaluating the resulting gaussian integral. This yields

$$P = \frac{q^2 A^2 \tau^2 \pi}{2m\omega\hbar} e^{-\omega^2 \tau^2 /2}. \tag{12.56}$$

Solution 3.5.   We can analyze this problem using the method of partial waves. A good reference is Cohen-Tanoudji, Volume 2.

a) A resonance is associated with a certain partial wave, and has the angular momentum quantum number l of that partial wave, whose angular dependence is described by a Legendre polynomial, P_l. The question tells us that the contribution from the resonance is nonzero everywhere except at $\theta = \pi/2$. The only Legendre polynomial satisfying this condition is P_1, and so we deduce that the angular momentum of the resonance is $J = 1$.

b) We need three results from the method of partial waves:

$$f_k(\theta) = \frac{1}{k} \sum_{l=0}^{\infty} (2l+1) e^{i\delta_l} \sin \delta_l \, P_l(\cos \theta), \tag{12.57}$$

$$\sigma(\theta) = |f_k(\theta)|^2, \tag{12.58}$$

$$\sigma_{tot} = \frac{4\pi}{k^2} \sum_{l=0}^{\infty} (2l+1) \sin^2 \delta_l. \tag{12.59}$$

Far off resonance, the cross-section is isotropic, so only the $l = 0$ term is present in $f(\theta)$. On resonance we also pick up the $l = 1$

term. Thus we can simplify the expression for the total cross-section at resonance (12.59):

$$\sigma_R = \frac{4\pi}{k^2}(\sin^2 \delta_0 + 3), \qquad (12.60)$$

where we have used $\delta_1 = \pi/2$, which corresponds to the peak of the resonance. We are given $\sigma_{tot} = \sigma_R$ and $k^2 = k_R^2$ on resonance, so we can solve for $\sin^2 \delta_0$ there:

$$\sin^2 \delta_0 = \left(\frac{k_R^2 \sigma_R}{4\pi} - 3 \right) \approx \frac{1}{2}. \qquad (12.61)$$

We want to find the differential cross-section at $\theta = \pi$, so we write out the two relevant terms of $f(\theta)$ and square their sum (being careful because $f(\theta)$ is complex), and insert the above expression for $\sin^2 \delta_0$. In the end, we find that

$$\sigma_R(\theta) = \frac{1}{k_R^2}[9 - 5\sin^2 \delta_0] \approx 6.5 \times 10^{-28} \text{cm}^2. \qquad (12.62)$$

Solution 3.6.   a) The Schrödinger equation for the two-component wavefunction Ψ of an electron at rest in a uniform magnetic field is

$$\frac{g_s \mu_B}{\hbar} \mathbf{S} \cdot \mathbf{B} \cdot \Psi = i\hbar \frac{\partial \Psi}{\partial t}. \qquad (12.63)$$

In this equation, the Bohr magneton is $\mu_B = e\hbar/2m_e c$, m_e is the electron mass, and g is the electron's gyromagnetic ratio. The spin is $\mathbf{S} = \hbar \boldsymbol{\sigma}/2$, where $\boldsymbol{\sigma}$ is the vector of Pauli spin matrices, presented here for ease of reference:

$$\sigma_x = \begin{pmatrix} 0 & 1 \\ 1 & 0 \end{pmatrix}, \quad \sigma_y = \begin{pmatrix} 0 & -i \\ i & 0 \end{pmatrix}, \quad \sigma_z = \begin{pmatrix} 1 & 0 \\ 0 & -1 \end{pmatrix}. \qquad (12.64)$$

If we write $\mathbf{B} = B_0 \hat{z}$ then the eigenstates are

$$\Psi_\uparrow(t) = \begin{pmatrix} 1 \\ 0 \end{pmatrix} e^{-i\Omega t} \quad \text{and} \quad \Psi_\downarrow(t) = \begin{pmatrix} 0 \\ 1 \end{pmatrix} e^{i\Omega t}, \qquad (12.65)$$

where $\Omega \equiv \mu_B B_0 / \hbar$.

Initially the electron has its spin pointing in the x-direction. This means that at $t = 0$ the wavefunction $\Psi(t)$ must be an eigenstate of the σ_x matrix, namely

$$\Psi(t = 0) = \frac{1}{\sqrt{2}} \begin{pmatrix} 1 \\ 1 \end{pmatrix}. \tag{12.66}$$

Alternatively we can write this in terms of the eigenstates (12.65). Then for arbitrary time the wavefunction is given by

$$\Psi(t) = \frac{1}{\sqrt{2}} (\Psi_\uparrow(t) + \Psi_\downarrow(t)) = \frac{1}{\sqrt{2}} \begin{pmatrix} e^{-i\Omega t} \\ e^{i\Omega t} \end{pmatrix}. \tag{12.67}$$

We can now calculate the probability of finding the spin in the x-direction at time t :

$$
\begin{aligned}
\langle S_x \rangle &= \frac{\hbar}{2} \langle \sigma_x \rangle = \frac{\hbar}{4} \begin{pmatrix} e^{i\Omega t} & e^{-i\Omega t} \end{pmatrix} \begin{pmatrix} 0 & 1 \\ 1 & 0 \end{pmatrix} \begin{pmatrix} e^{-i\Omega t} \\ e^{i\Omega t} \end{pmatrix} \\
&= \frac{\hbar}{2} \cos 2\Omega t.
\end{aligned} \tag{12.68}
$$

We can also find the other components of $\langle \mathbf{S}(t) \rangle$:

$$\langle S_y \rangle = \frac{\hbar}{2} \sin 2\Omega t \quad \text{and} \quad \langle S_z \rangle = 0. \tag{12.69}$$

So we see that the spin precesses around the magnetic field with an angular precession frequency of 2Ω.

b) When an additional time-dependent magnetic field $\mathbf{B}_1$ is applied, it is tempting to try to use perturbation theory. However the question makes no mention of $\mathbf{B}_1$ being "weak." Instead, we find an exact solution.

Our first step is to express the interaction term in a useful form:

$$\mathbf{B} \cdot \boldsymbol{\sigma} = \begin{pmatrix} B_0 & \frac{1}{2} B_1 e^{-i\omega t} \\ \frac{1}{2} B_1 e^{i\omega t} & -B_0 \end{pmatrix}. \tag{12.70}$$

Now we substitute this into the Schrödinger equation (12.63) and look for two solutions of the form

$$\Psi(t) = \begin{pmatrix} a \, e^{i\omega_a t} \\ b \, e^{i\omega_b t} \end{pmatrix}. \tag{12.71}$$

This form can be motivated as follows. We can see that the wave-function cannot have a simple exponential time dependence, as the Schrödinger equation couples the two components in a nontrivial, time-dependent way. However the time dependence of the *interaction* is simply that of an exponential, and we can hope to find a solution which is some combination of exponentials of different frequencies. In fact this turns out to be the case, as we will see. If we insert our would-be wavefunction (12.71) into the Schrödinger equation (12.63), we obtain the following linear equations in a and b:

$$\mu_B \left(B_0 e^{i\omega_a t} a + \frac{1}{2} B_1 e^{i(\omega_b - \omega)t} b \right) = -\hbar \omega_a e^{i\omega_a t} a, \tag{12.72}$$

$$\mu_B \left(\frac{1}{2} B_1 e^{i(\omega_a + \omega)t} a - B_0 e^{i\omega_b t} b \right) = -\hbar \omega_b e^{i\omega_b t} b. \tag{12.73}$$

Our first condition is that within each equation, the time dependence of all the terms should be the same, which requires

$$\omega_b - \omega_a = \omega. \tag{12.74}$$

Before we derive the other conditions, we note that we can set $a = 1$ without loss of generality. Further, for sake of clarity let us define $\beta = \mu_B B_1 / 2\hbar$ After we cancel the common exponential time dependence, our equations now reduce to the simple form

$$-\omega_a = \Omega + \beta b, \tag{12.75}$$

$$-\omega_b = -\Omega + \frac{\beta}{b}. \tag{12.76}$$

If we combine these with equation (12.74), we obtain a quadratic equation for b:

$$\omega_b - \omega_a = \omega = 2\Omega + \beta \left(b - \frac{1}{b} \right), \tag{12.77}$$

or equivalently,

$$b^2 - b \left(\frac{\omega}{\beta} - \frac{2\Omega}{\beta} \right) - 1 = 0. \tag{12.78}$$

We can solve this to find the two possible values of b, and then, from equations (12.75) and (12.76), obtain the frequencies of the components:

$$b_\pm = \frac{\omega - 2\Omega}{2\beta} \pm \frac{\Delta}{\beta}, \tag{12.79}$$

$$\omega_a^{\pm} = -\frac{\omega}{2} \mp \Delta, \quad \text{and} \tag{12.80}$$

$$\omega_b^{\pm} = \frac{\omega}{2} \mp \Delta, \tag{12.81}$$

where we have defined

$$\Delta \equiv \sqrt{\beta^2 + \left(\frac{\omega}{2} - \Omega\right)^2}. \tag{12.82}$$

The unnormalized eigenvectors of the hamiltonian (12.63) with $\mathbf{B} = \mathbf{B}_0 + \mathbf{B}_1$ are then

$$\begin{pmatrix} e^{i\omega_a^+ t} \\ b_+ e^{i\omega_b^+ t} \end{pmatrix} \quad \text{and} \quad \begin{pmatrix} e^{i\omega_a^- t} \\ b_- e^{i\omega_b^- t} \end{pmatrix}. \tag{12.83}$$

Initially we have $\langle S_z \rangle = +\hbar/2$, or

$$\Psi(t = 0) = \begin{pmatrix} 1 \\ 0 \end{pmatrix}. \tag{12.84}$$

At time t this will have evolved into a linear superposition of the eigenvectors (12.83):

$$\Psi(t) = p \begin{pmatrix} e^{i\omega_a^+ t} \\ b_+ e^{i\omega_b^+ t} \end{pmatrix} + q \begin{pmatrix} e^{i\omega_a^- t} \\ b_- e^{i\omega_b^- t} \end{pmatrix}, \tag{12.85}$$

for some constants p and q. Our initial condition (12.84) gives us $p+q = 1$, and $pb_+ + qb_- = 0$. (Note that since Ψ is normalized at $t = 0$, it remains normalized for all time.) These can be solved to give

$$p = -\frac{\beta b_-}{2\Delta}, \quad \text{and} \quad q = \frac{\beta b_+}{2\Delta}. \tag{12.86}$$

After a time t, the probability that the electron is in a state with $\langle S_z \rangle = -\hbar/2$ is the modulus squared of the lower component of $\Psi(t)$,

$$P(t) = \left| p \, b_+ e^{i\omega_b^+ t} + q \, b_- e^{i\omega_b^- t} \right|^2 = \left| \frac{\beta b_1 b_+}{2\Delta} \right|^2 \left| e^{i\Delta t} - e^{-i\Delta t} \right|^2$$

$$= \frac{\beta^2}{\Delta^2} \sin^2 \Delta t. \tag{12.87}$$

If the oscillating magnetic field has an angular frequency $\omega = 2\Omega$ then it is "at resonance" and $P(t) = \sin^2 \beta t$. This precession of the expectation value of the spin with angular frequency 2β is called Rabi precession, and 2β is called the Rabi flopping frequency.

Solution 3.7.   a) The system wavefunction factors into spatial and isospin wavefunctions. The spatial part contributes equally to the $I = 3/2$ and $I = 1/2$ cross-sections, so for calculating the ratio of cross-sections we can concentrate entirely on the isospin part. The total isospin operator is $\mathbf{I} = \mathbf{I}^{(\pi)} + \mathbf{I}^{(N)}$. Squaring this expression and rearranging terms gives an expression for the isospin portion of the matrix element:

$$\langle \mathbf{I}^{(\pi)} \cdot \mathbf{I}^{(N)} \rangle = \frac{1}{2} \left[I(I+1) - I^{(\pi)}(I^{(\pi)} + 1) - I^{(N)}(I^{(N)} + 1) \right], \quad (12.88)$$

where we have computed the expectation value in a state of definite total I. The pion and nucleon have isospins 1 and 1/2, respectively, so for the total isospin $I = 3/2$ case $\langle \mathbf{I}^{(\pi)} \cdot \mathbf{I}^{(N)} \rangle = 1/2$, and for the $I = 1/2$ case $\langle \mathbf{I}^{(\pi)} \cdot \mathbf{I}^{(N)} \rangle = -1$. To first order in perturbation theory, the cross-section depends on the square of the matrix element of the potential. Thus the ratio of the $I = 3/2$ to the $I = 1/2$ cross-sections is

$$\frac{\sigma_{I=3/2}}{\sigma_{I=1/2}} = \left(\frac{1/2}{-1} \right)^2 = 1/4. \quad (12.89)$$

b) In the Born approximation, the differential cross-section is given by

$$\frac{d\sigma}{d\Omega} = |f|^2, \quad (12.90)$$

where

$$f = \frac{m}{2\pi\hbar^2} \tilde{U}(\mathbf{q}) \langle \mathbf{I}^{(\pi)} \cdot \mathbf{I}^{(N)} \rangle, \quad (12.91)$$

$\mathbf{q} = \mathbf{k}_i - \mathbf{k}_f$ is the momentum transfer, and $\tilde{U}(\mathbf{q})$ is the Fourier transform of the potential. We are interested in the potential

$$U(r) = \frac{a^2}{4\pi} \frac{e^{-\mu r}}{r}, \tag{12.92}$$

which has the Fourier transform

$$\tilde{U}(\mathbf{q}) = \frac{a^2}{4\pi} \int d^3x \frac{e^{-\mu r}}{r} e^{i\mathbf{q}\cdot\mathbf{r}}. \tag{12.93}$$

Writing $\mathbf{q} \cdot \mathbf{r} = qr \cos\theta$ (where θ is measured with respect to the direction of $\mathbf{q}$), this becomes

$$
\begin{aligned}
\tilde{U}(\mathbf{q}) &= \frac{a^2}{4\pi} \int dr\, r^2\, d\phi\, d(\cos\theta) \frac{e^{-\mu r}}{r} e^{iqr\cos\theta} \\
&= \frac{a^2}{2iq} \int_0^\infty dr \left(e^{-r(\mu-iq)} - e^{-r(\mu+iq)} \right) \\
&= \frac{a^2}{\mu^2 + q^2}. \tag{12.94}
\end{aligned}
$$

We note now that $\tilde{U}$ depends only on the magnitude of $\mathbf{q}$. To find the total cross-section, we integrate equation (12.90) over the unit sphere, which gives a factor of 4π. So

$$
\begin{aligned}
\sigma &= \frac{m^2 a^4}{\pi \hbar^4 (\mu^2 + q^2)^2} \left| \langle \mathbf{I}^{(\pi)} \cdot \mathbf{I}^{(N)} \rangle \right|^2 \\
&\equiv k \left| \langle \mathbf{I}^{(\pi)} \cdot \mathbf{I}^{(N)} \rangle \right|^2, \tag{12.95}
\end{aligned}
$$

where we have introduced a new constant k, as indicated.

In order to calculate $\langle I^{(\pi)} \cdot I^{(N)} \rangle$ using equation (12.88), we must first write the initial and final states in a basis of total isospin I eigenstates. Written in such an $|I\, I_Z\rangle$ basis, the relevant states are

$$|\pi^+ p\rangle = |\tfrac{3}{2}\, \tfrac{3}{2}\rangle, \tag{12.96}$$

$$|\pi^- p\rangle = \frac{1}{\sqrt{3}} |\tfrac{3}{2}\, \tfrac{-1}{2}\rangle - \sqrt{\frac{2}{3}} |\tfrac{1}{2}\, \tfrac{-1}{2}\rangle, \tag{12.97}$$

$$|\pi^0 n\rangle = \sqrt{\frac{2}{3}} |\tfrac{3}{2}\, \tfrac{-1}{2}\rangle + \frac{1}{\sqrt{3}} |\tfrac{1}{2}\, \tfrac{-1}{2}\rangle, \tag{12.98}$$

where a table of Clebsch-Gordon coefficients has been consulted. For the reaction $\pi^+ + p \rightarrow \pi^+ + p$ we evaluate equation (12.95) for the wavefunction (12.96), using the results from part (a), to find the total cross-section:

$$\sigma = k \left| \frac{1}{2} \right|^2 = \frac{k}{4}. \tag{12.99}$$

Similarly, for $\pi^- + p \rightarrow \pi^- + p$ the total cross-section is

$$\sigma = k \left| \frac{1}{3} \cdot \frac{1}{2} + \frac{2}{3} \cdot (-1) \right|^2 = \frac{k}{4}, \tag{12.100}$$

and for $\pi^- + p \rightarrow \pi^0 + n$ the total cross-section is

$$\sigma = k \left| \frac{\sqrt{2}}{3} \cdot \frac{1}{2} - \frac{\sqrt{2}}{3} \cdot (-1) \right|^2 = \frac{k}{2}. \tag{12.101}$$

Solution 3.8. Within the nth region the particle moves in a constant potential, so we may write its wavefunction as a sum of exponentials:

$$\psi_n = C_n e^{ik_n x} + D_n e^{-ik_n x}. \tag{12.102}$$

The two terms represent the right- and left-moving components, respectively. In the expression above, the wavevector k_n is given by

$$\frac{\hbar^2 k_n^2}{2m} = (E - V_n), \tag{12.103}$$

from the time-independent Schrödinger equation in the nth region. We solve for k_n to find $k_n = n\alpha$. The boundary conditions on (12.102) follow from the requirement of continuity of the wavefunction and its derivative throughout space:

$$\begin{aligned} \psi_n(x = x_n) &= \psi_{n+1}(x = x_n), \\ \left. \frac{\partial \psi_n}{\partial x} \right|_{x=x_n} &= \left. \frac{\partial \psi_{n+1}}{\partial x} \right|_{x=x_n}, \end{aligned} \tag{12.104}$$

where $x_n = (n - 1)\pi/\alpha$. If we use the form of the wavefunction given in equation (12.102), we find the intermediate boundary conditions:

$$
\begin{aligned}
C_n + D_n &= (C_{n+1} + D_{n+1})e^{i(n^2-1)\pi} = (C_{n+1} + D_{n+1})(-1)^{n+1}, \\
n(C_n - D_n) &= (n+1)(C_{n+1} - D_{n+1})(-1)^{n+1}.
\end{aligned}
\tag{12.105}
$$

We wish to relate the coefficients C_1 and D_1 to C_N and D_N. We can see immediately from the equations above that the relation has the form:

$$
\begin{aligned}
C_1 + D_1 &= (C_N + D_N)(-1)^m, \\
C_1 - D_1 &= N(C_N - D_N)(-1)^m,
\end{aligned}
\tag{12.106}
$$

where m is an integer. In fact, $m = \sum_{n=2}^{n=N} n$, but the important point is not its value, but the fact that m is the same in both of the equations above. For a particle entering from the left, we set $D_N = 0$ and solve for the transmission coefficient T_L, given by the expression

$$
T_L = \left|\frac{C_N}{C_1}\right|^2 \frac{k_N}{k_1}.
\tag{12.107}
$$

(For a proof of this relation, see any introductory quantum mechanics text.) We find

$$
T_L = \frac{4N}{(N+1)^2}.
\tag{12.108}
$$

Conversely, if the particle enters from the right, we set $C_1 = 0$ and solve for

$$
T_R = \left|\frac{D_1}{D_N}\right|^2 \frac{k_1}{k_N} = \frac{4N}{(N+1)^2}.
\tag{12.109}
$$

We have discovered that the transmission coefficient is the same for particles incident from either side of this potential, a special case of a more general result.

Solution 3.9. a) The Schrödinger equation for a neutron at rest in a uniform magnetic field **B** is

$$
\mu\mathbf{B} \cdot \boldsymbol{\sigma}|\Psi(t)\rangle = i\hbar\frac{\partial}{\partial t}|\Psi(t)\rangle,
\tag{12.110}
$$

where $\boldsymbol{\sigma}$ is the vector of Pauli spin matrices and $|\Psi(t)\rangle$ is a two-component wavefunction. Here μ is the magnetic dipole moment of the neutron, and we do not need to know its value except that it is negative. Suppose we write the magnetic field as $\mathbf{B} = B_0\hat{\mathbf{n}}$. We will look for states of definite energy, which we can write as $|\Psi(t)\rangle = |\Psi\rangle\exp(-iEt/\hbar)$. Then $\mu\mathbf{B}\cdot\boldsymbol{\sigma}|\Psi(t)\rangle = E|\Psi(t)\rangle$. We expect to find two eigenstates, one with its spin aligned with the magnetic field and the other with its spin anti-aligned. Unfortunately, since $\hat{\mathbf{n}}$ is pointing in an arbitrary direction, it is not simple to down these eigenstates. We must find the eigenvectors of the matrix $\hat{\mathbf{n}}\cdot\boldsymbol{\sigma}$, which we denote by two-component column vectors. The eigenvalue equation is

$$\begin{pmatrix} n_3 & n_1 - in_2 \\ n_1 + in_2 & -n_3 \end{pmatrix}\begin{pmatrix} a \\ b \end{pmatrix} = \lambda\begin{pmatrix} a \\ b \end{pmatrix}. \qquad (12.111)$$

For a given eigenvalue λ, the energy is $E = \lambda\mu B_0$. For a nontrivial solution we require that $\det[\hat{\mathbf{n}}\cdot\boldsymbol{\sigma} - \lambda\mathbf{I}] = 0$, which we can rewrite as

$$-(n_3 - \lambda)(n_3 + \lambda) - (n_1 - in_2)(n_1 + in_2) = \lambda^2 - \hat{\mathbf{n}}^2 = 0. \quad (12.112)$$

Therefore we find $\lambda = \pm 1$.

The energy eigenstates are the corresponding solutions of (12.111):

$$\begin{pmatrix} n_3 \mp 1 & n_1 - in_2 \\ n_1 + in_2 & \mp 1 - n_3 \end{pmatrix}\begin{pmatrix} a \\ b \end{pmatrix} = 0. \qquad (12.113)$$

One possible solution is

$$\begin{pmatrix} a \\ b \end{pmatrix} = \begin{pmatrix} n_1 - in_2 \\ \pm 1 - n_3 \end{pmatrix}. \qquad (12.114)$$

However, this choice is not unique, and in particular we can multiply by an overall phase and still have a solution to (12.113). This will turn out to be important, so for the moment we will explicitly include an arbitrary phase $e^{i\alpha}$.

For a particular direction of $\mathbf{B}$, we can write $\hat{\mathbf{n}}$ in terms of the angles θ and ϕ: $n_1 = \sin\theta\cos\phi$, $n_2 = \sin\theta\sin\phi$, and $n_3 = \cos\theta$. Therefore the eigenstates are

$$|\Psi_\pm\rangle = \frac{1}{N_\pm}\begin{pmatrix} \sin\theta[\cos\phi - i\sin\phi] \\ \pm 1 - \cos\theta \end{pmatrix}e^{i\alpha_\pm}. \qquad (12.115)$$

The prefactors $N_\pm$ are real normalization constants, which can be found after a small amount of algebra and an application of the trigonometric half-angle identities. We can write the normalized eigenstates as

$$|\Psi_+\rangle = \begin{pmatrix} e^{-i\phi} \cos\frac{\theta}{2} \\ \sin\frac{\theta}{2} \end{pmatrix} e^{i\alpha_+}, \tag{12.116}$$

$$|\Psi_-\rangle = \begin{pmatrix} e^{-i\phi} \sin\frac{\theta}{2} \\ -\cos\frac{\theta}{2} \end{pmatrix} e^{i\alpha_-}. \tag{12.117}$$

We now come to the subtle question of fixing the phases $\alpha_\pm$. Suppose we were to take $\alpha_+ = 0$. Then for $\theta = 0$ we would get

$$|\Psi_+\rangle = \begin{pmatrix} e^{-i\phi} \\ 0 \end{pmatrix}, \tag{12.118}$$

which is not well-defined, as the angle ϕ can have any value. In particular, our wavefunction is not differentiable at $\theta = 0$. (For a discussion of this point, see Jordan.) The important issue here is that in the limit $\theta \to 0$, our eigenstates should not depend on ϕ. We can achieve this by taking $\alpha_+ = \phi$, and $\alpha_- = 0$. We now have our final result for the eigenstates:

$$|\Psi_+\rangle = \begin{pmatrix} \cos\frac{\theta}{2} \\ e^{i\phi} \sin\frac{\theta}{2} \end{pmatrix}, \tag{12.119}$$

$$|\Psi_-\rangle = \begin{pmatrix} e^{-i\phi} \sin\frac{\theta}{2} \\ -\cos\frac{\theta}{2} \end{pmatrix}. \tag{12.120}$$

These two states correspond to neutrons with spins respectively aligned and anti-aligned with the direction of the magnetic field, and with energy $\mp E_0$, for $E_0 = |\mu B_0|$. Because the neutron magnetic dipole moment μ is negative, the state with lower energy will be the one for which the spin is aligned with the magnetic field; i.e., our ground state is $|\Psi_+\rangle$.

b) Suppose the neutron beams take a time T to pass through the magnetic fields. In the adiabatic approximation, we consider the limit $T \to \infty$, and we can use $1/T$ as the small parameter of a perturbation expansion.

Since the direction of the field varies slowly, the neutron will be very nearly in the ground state $|\Psi_+\rangle$ of the local magnetic field. However $|\Psi_+\rangle$ is a function of ϕ, which is itself a function of t, and so $|\Psi_+\rangle e^{iE_0 t/\hbar}$ no longer satisfies the time-dependent Schrödinger equation. Because of this, we must expand the wavefunction $|\Psi(t)\rangle$ in terms of the basis $|\Psi_+\rangle$ and $|\Psi_-\rangle$, with coefficients that are functions of time:

$$|\Psi(t)\rangle = a(t)|\Psi_+\rangle + b(t)|\Psi_-\rangle. \tag{12.121}$$

The great simplification that occurs for large T is that the coefficient $b(t)$ must be small, of order $1/T$. If $|\Psi(t)\rangle$ is normalized to unity, it follows that $|a|^2 = 1 - |b|^2 = 1 - \mathcal{O}(1/T^2)$. From now on we will drop terms that are smaller than $1/T$. Therefore $|a|^2 = 1$, and we will later write $a(t)$ in the general form $a(t) = e^{i\gamma(t)/\hbar}$, where $\gamma(t)$ is a real function.

We can now substitute the wavefunction (12.121) into the Schrödinger equation (12.110):

$$i\hbar\frac{\partial}{\partial t}|\Psi(t)\rangle = \mu\mathbf{B}\cdot\boldsymbol{\sigma}|\Psi(t)\rangle = -E_0 a(t)|\Psi_+\rangle + E_0 b(t)|\Psi_-\rangle. \tag{12.122}$$

Consider the left-hand side:

$$\frac{\partial}{\partial t}|\Psi(t)\rangle \equiv \frac{da}{dt}|\Psi_+\rangle + a\frac{\partial}{\partial t}|\Psi_+\rangle + \frac{db}{dt}|\Psi_-\rangle + b\frac{\partial}{\partial t}|\Psi_-\rangle. \tag{12.123}$$

If we multiply throughout by $\langle\Psi_+|$, the only terms that contribute on the right-hand side are the first two; the third term is clearly orthogonal to $\langle\Psi_+|$, and the fourth term is already $\mathcal{O}(1/T^2)$, since

$$b\frac{\partial}{\partial t}|\Psi_-\rangle = b\frac{d\phi}{dt}\frac{\partial}{\partial\phi}|\Psi_-\rangle \tag{12.124}$$

and $d\phi/dt = 2\pi/T$.

Thus after multiplication by $\langle\Psi_+|$, (12.122) reduces to

$$i\hbar\frac{da}{dt} + a(t)\langle\Psi_+|i\hbar\frac{\partial}{\partial t}|\Psi_+\rangle = -E_0 a(t). \tag{12.125}$$

Using the form $a(t) = e^{i\gamma(t)/\hbar}$ and canceling a common phase, we get

$$
\begin{aligned}
\frac{d\gamma}{dt} - E_0 &= \langle \Psi_+ | i\hbar \frac{\partial}{\partial t} | \Psi_+ \rangle \\
&= -\hbar(\cos\tfrac{\theta}{2}, \sin\tfrac{\theta}{2}e^{-i\phi})\frac{d\phi}{dt}\begin{pmatrix} 0 \\ \sin\tfrac{\theta}{2}e^{i\phi} \end{pmatrix} \\
&= -\hbar(\sin^2\tfrac{\theta}{2})\frac{d\phi}{dt}.
\end{aligned}
\tag{12.126}
$$

We can now integrate from $t = 0$ to $t = T$, during which time ϕ goes from 0 to 2π:

$$
\begin{aligned}
\frac{1}{\hbar}\int_0^T dt\left(\frac{d\gamma}{dt} - E_0\right) &= \frac{1}{\hbar}\left(\gamma(T) - \gamma(0) - E_0 T\right) \tag{12.127} \\
&= -\sin^2\tfrac{\theta}{2}\int_0^{2\pi} d\phi = -2\pi\sin^2\tfrac{\theta}{2}.
\end{aligned}
$$

Now $[\gamma(T) - \gamma(0)]/\hbar$ is the phase change of the beam that went through the changing magnetic field, and $E_0 T/\hbar$ is the phase change of the other beam. Hence the difference $\Delta = -2\pi\sin^2\theta/2$ is the phase shift between the two beams, which is the quantity we are after. When we recombine the beams, the total wavefunction is

$$
|\Psi_T\rangle = \left(e^{i[\gamma(T) - \gamma(0)]/\hbar} + e^{iE_0 T/\hbar}\right)|\Psi_+\rangle + b(t)|\Psi_-\rangle,
\tag{12.128}
$$

up to an overall normalization. The intensity of the beam I is proportional to

$$
\langle\Psi_T|\Psi_T\rangle = |1 + e^{i\Delta}|^2 + \mathcal{O}\left(\frac{1}{T^2}\right).
\tag{12.129}
$$

We can normalize this expression so that when $\Delta = 0$, $I = 1$. Thus we get our final result:

$$
I = \cos^2\tfrac{\Delta}{2} = \cos^2\left(\pi\sin^2\tfrac{\theta}{2}\right).
\tag{12.130}
$$

This problem illustrates a simple example of Berry phases.

Solution 3.10. First we choose units such that $\hbar = 1$. In the orbital ground state, the orbital angular momentum is zero, so the relevant part of the Hamiltonian is

$$H = (\alpha \mathbf{S}_1 + \beta \mathbf{S}_2) \cdot \mathbf{B} + J \mathbf{S}_1 \cdot \mathbf{S}_2. \qquad (12.131)$$

Let us choose the z-axis to be parallel to the uniform magnetic field, $\mathbf{B} = B\hat{z}$. Then

$$H = (\alpha S_{1z} + \beta S_{2z})B + J \left[S_{1z}S_{2z} + \frac{1}{2}(S_1^+ S_2^- + S_1^- S_2^+) \right], \qquad (12.132)$$

where $S^{\pm} = S_x \pm iS_y$. For two particles with spin, we usually describe the spin part of the wavefunction in either the basis of states given by $|S_1, S_2, S_{1z}, S_{2z}\rangle$, or $|S, S_z, S_1, S_2\rangle$, where we define $\mathbf{S} = \mathbf{S}_1 + \mathbf{S}_2$. H is not diagonal in either of these bases, for general α, β, and J. However, since the proton and electron are spin $1/2$ particles, we have only a small number of basis vectors and we can solve the problem by brute force. Let us choose the first basis suggested above, writing the basis vectors as

$$|\psi_1\rangle = |\uparrow\uparrow\rangle, \quad |\psi_2\rangle = |\uparrow\downarrow\rangle, \quad |\psi_3\rangle = |\downarrow\uparrow\rangle, \quad |\psi_4\rangle = |\downarrow\downarrow\rangle, \qquad (12.133)$$

where the first arrow represents the S_{1z}, and the second arrow represents the S_{2z}.

If we form the 4×4 matrix $\langle \psi_i | H | \psi_j \rangle$ then, by definition, its eigenvalues are the energy eigenvalues and its eigenvectors are the eigenstates of the system. We find that the matrix elements are given by:

$$\begin{pmatrix} \frac{(\alpha+\beta)B}{2} + \frac{J}{4} & 0 & 0 & 0 \\ 0 & \frac{(\alpha-\beta)B}{2} - \frac{J}{4} & \frac{J}{2} & 0 \\ 0 & \frac{J}{2} & -\frac{(\alpha-\beta)B}{2} - \frac{J}{4} & 0 \\ 0 & 0 & 0 & -\frac{(\alpha+\beta)B}{2} + \frac{J}{4} \end{pmatrix}. \qquad (12.134)$$

Let us denote the four eigenvectors of this matrix as ϕ_a, ϕ_b, ϕ_c and ϕ_d. Since H is in block-diagonal form, we can immediately write down two eigenvectors and their corresponding eigenvalues:

$$|\phi_a\rangle = |\uparrow\uparrow\rangle \quad \text{with} \quad E_a = +\frac{B}{2}(\alpha + \beta) + \frac{J}{4}, \qquad (12.135)$$

$$|\phi_b\rangle = |\downarrow\downarrow\rangle \quad \text{with} \quad E_b = -\frac{B}{2}(\alpha + \beta) + \frac{J}{4}. \qquad (12.136)$$

To find the other two eigenenergies and eigenstates we need to diagonalize the submatrix

$$\mathbf{A} = \begin{pmatrix} \frac{B}{2}(\alpha - \beta) - \frac{J}{4} & \frac{J}{2} \\ \frac{J}{2} & -\frac{B}{2}(\alpha - \beta) - \frac{J}{4} \end{pmatrix}. \qquad (12.137)$$

The eigenvalues λ of $\mathbf{A}$ are given by the quadratic equation $\det[\mathbf{A} - \lambda\mathbf{I}] = 0$. Solving this equation for the two eigenvalues yields

$$E_c = -\frac{J}{4} + \frac{1}{2}k, \qquad (12.138)$$

$$E_d = -\frac{J}{4} - \frac{1}{2}k, \qquad (12.139)$$

where for simplicity we have defined

$$k = \sqrt{J^2 + B^2(\alpha - \beta)^2}. \qquad (12.140)$$

We sketch the energy splittings as a function of B in Figure 12.3.

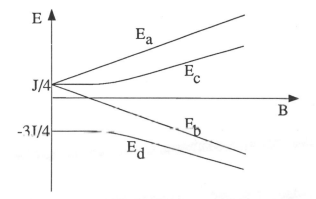

Figure 12.3.

b) Two of the eigenvectors, $\phi_a = |\uparrow\uparrow\rangle$ and $\phi_b = |\downarrow\downarrow\rangle|$ are given above. The other two are the normalized eigenvectors of the submatrix **A**:

$$|\phi_c\rangle \;=\; \frac{1}{\sqrt{N}}\left\{|\uparrow\downarrow\rangle + \frac{1}{J}\,[k - B(\alpha - \beta)]\,|\downarrow\uparrow\rangle\right\}, \qquad (12.141)$$

$$|\phi_d\rangle \;=\; \frac{1}{\sqrt{N}}\left\{\frac{1}{J}\,[-k + B(\alpha - \beta)]\,|\uparrow\downarrow\rangle + |\downarrow\uparrow\rangle\right\}, \qquad (12.142)$$

where

$$N = 1 + \frac{1}{J^2}\,(k - B(\alpha - \beta))^2. \qquad (12.143)$$

We note that for $B = 0$, the eigenvectors of **A** reduce to the basis $|S, S_z, S_1, S_2\rangle$, and for $J = 0$ they reduce to the basis $|S_1, S_2, S_{1z}, S_{2z}\rangle$.

Chapter 13

Thermodynamics &
Statistical
Mechanics—Solutions

Solution 4.1. a) We first write down the number of available states using the binomial distribution:

$$g = \frac{N!}{n_0! \, n_1!},$$ (13.1)

and then use Stirling's approximation to express the entropy, given by $S = k \ln g$, as

$$S = k[N \ln N - n_1 \ln n_1 - n_0 \ln n_0],$$ (13.2)

where k is the Boltzmann constant. We know that n_0 and n_1 satisfy the conditions

$$N = n_0 \mid n_1 \quad \text{and} \quad U = n_1 E.$$ (13.3)

Solving these for n_0 and n_1 and substituting into our equation for S gives

$$S = k \left[N \ln N - \frac{U}{E} \ln \frac{U}{E} - \left(N - \frac{U}{E} \right) \ln \left(N - \frac{U}{E} \right) \right].$$ (13.4)

b) For a constant number of particles, the temperature can be found from

$$\frac{1}{T} = \left(\frac{\partial S}{\partial U}\right)_N. \tag{13.5}$$

Using the expression for entropy (13.4) gives us the temperature:

$$T = \frac{E}{k \ln(EN/U - 1)}. \tag{13.6}$$

To find the range of n_0 for which $T < 0$, we switch variables from U and N to n_0 and n_1, using equation (13.3). This gives

$$\frac{1}{T} = \frac{k}{E}(\ln n_0 - \ln n_1). \tag{13.7}$$

We can see that $T < 0$ when $n_0 < n_1$, so that the temperature is negative for $0 < n_0 < N/2$.

c) As the systems approach thermal equilibrium, ΔS_{total} must be greater than zero. We know that in each system, $\Delta S = \Delta Q/T$. Suppose system 1 has $T < 0$ and system 2 has $T > 0$. If heat flows to system 1 from system 2, then $\Delta Q_1 > 0$ and $\Delta Q_2 < 0$, implying that $\Delta S < 0$ in both subsystems. This cannot be true. Conversely, if heat flows from system 1 to system 2, $\Delta S > 0$ in both systems, which is allowed. Thus heat must flow from the system with negative temperature to the system with positive temperature. This makes sense, because most of the energy is in the system with negative temperature.

Solution 4.2.   a) In order to calculate the specific heat of a classical system, it is necessary only to know the number of degrees of freedom of the system. The specific heat then follows from an application of the equipartition theorem.

For the case of a heteronuclear diatomic molecule, there are some subtleties to do with quantum mechanics that we cannot ignore. Classically the angular momentum vector is free to point in any direction with respect to the orientation of the molecule, and since the rotational energy is $\frac{1}{2}(I_1\omega_1^2 + I_2\omega_2^2 + I_3\omega_3^2)$, one might naively expect there to be three degrees of freedom. However, it is a consequence of quantum mechanics that a system *cannot* rotate about an axis of continuous symmetry, which in the case of a diatomic molecule means that there can be no component of angular momentum along the axis joining the two atoms (see Burcham). This microscopic constraint has a macroscopically observable effect, namely that the number of degrees of freedom of rotation is reduced from three to two. (Often the same result is "derived" classically, by arguing that the moment of inertia about the symmetry axis is zero.)

The classical average internal energy is given by

$$\langle E \rangle = (\# \text{ degrees of freedom}) \times \frac{1}{2}kT = kT, \qquad (13.8)$$

where k is Boltzmann's constant. Therefore the specific heat is given by

$$C = \frac{d\langle E \rangle}{dT} = k. \qquad (13.9)$$

b) The partition function $\mathcal{Z}$ is defined to be the sum of the Boltzmann factor $e^{-E/kT}$ over all distinct quantum states of the system. For a system with energy levels $E_j = (\hbar^2/2I)j(j+1)$, each having degeneracy $(2j+1)$, we find

$$\mathcal{Z} = \sum_{j=0}^{\infty}(2j+1)e^{-\beta E_j}, \qquad (13.10)$$

where $\beta \equiv 1/kT$. The average energy $\langle E \rangle$ is defined as

$$\langle E \rangle = \frac{\displaystyle\sum_{j=0}^{\infty} E_j(2j+1)e^{-\beta E_j}}{\displaystyle\sum_{j=0}^{\infty}(2j+1)e^{-\beta E_j}} - -\frac{\partial}{\partial \beta}\ln \mathcal{Z}. \qquad (13.11)$$

c) As $T \to 0$ ($\beta \to \infty$), the occupation numbers of the higher excited states will be heavily suppressed by the Boltzmann factor, and for sufficiently low temperature we can replace the infinite sums in equations (13.10) and (13.11) by finite sums over just the ground state and the first excited state. We get

$$\mathcal{Z} \approx 1 + 3e^{-\beta E_1}, \tag{13.12}$$

so that

$$\ln \mathcal{Z} \approx 3e^{-\beta E_1}, \tag{13.13}$$

where we have assumed that $3e^{-\beta E_1} \ll 1$ and used the small-x expansion $\ln(1 + x) \approx x$. Therefore, at low temperatures,

$$\langle E \rangle \approx -\frac{\partial}{\partial \beta} 3e^{-\beta E_1} = 3E_1 e^{-E_1/kT}. \tag{13.14}$$

We can differentiate this to find the low-temperature specific heat,

$$C = \frac{d\langle E \rangle}{dT} = \frac{3E_1^2}{kT^2} e^{-E_1/kT}. \tag{13.15}$$

We note that for small T, the exponential suppression dominates over the $1/T^2$ prefactor, and $C \to 0$ as $T \to 0$.

Our derivation is valid provided that the occupation number of the second level is far smaller than that of the first ($5e^{-\beta E_2} \ll 3e^{-\beta E_1}$), and that the occupation number of the first level is much smaller than unity ($3e^{-\beta E_1} \ll 1$). In fact both of these yield the same condition,

$$kT \ll \frac{\hbar^2}{I}. \tag{13.16}$$

d) In the limit $T \to \infty$ ($\beta \to 0$), many states will become heavily populated and contribute to the sums in the partition function (13.10) and the average energy (13.11). Under these circumstances it becomes legitimate to approximate the discrete sum by an integral, with vanishingly small error for large enough T. Therefore we can write

$$\mathcal{Z} \approx \int_0^\infty dj\, (2j + 1) \exp\left(-\frac{\beta \hbar^2}{2I} j(j + 1)\right). \tag{13.17}$$

We integrate to obtain

$$\mathcal{Z} \approx \frac{-2I}{\beta\hbar^2} \left[\exp\left(-\frac{\beta\hbar^2}{2I} j(j+1) \right) \right]_0^\infty = \frac{2I}{\beta\hbar^2}, \qquad (13.18)$$

so that

$$\ln \mathcal{Z} \approx -\ln\beta + \text{constant}. \qquad (13.19)$$

Thus we find for the average internal energy

$$\langle E \rangle = -\frac{\partial}{\partial\beta} \ln \mathcal{Z} \approx \frac{1}{\beta} = kT, \qquad (13.20)$$

which we recognize as the classical result. Therefore the specific heat at high temperatures is simply

$$C = \frac{d\langle E \rangle}{dT} \approx k. \qquad (13.21)$$

This result is valid provided the Boltzmann factor is large up to values of j much greater than unity, that is $\beta E_1 \ll 1$, or $kT \gg \hbar^2/I$. There is, of course, another limit to the validity of our expression, namely that the temperature must not be so high that the molecule dissociates.

———

Solution 4.3. After N jumps the impurity will have made n_+ jumps to the right, and $n_- = N - n_+$ jumps to the left. Consequently, the impurity will have moved a distance

$$d = a(n_+ - n_-) \equiv as. \qquad (13.22)$$

There is an equal probability that the impurity jumps right or left. Therefore, the probability that the atom makes n_+ jumps to the right out of a total of N jumps is

$$P(n_+) = \binom{N}{n_+} = \frac{N!}{n_+! \, n_-!}, \qquad (13.23)$$

or, using the definition of s given in equation (13.22),

$$P(s) = \frac{N!}{[(N+s)/2]!\,[(N-s)/2]!}. \tag{13.24}$$

This equation gives the unnormalized probability that the impurity moves a distances $d = as$. In the limit of large N, we can express the probability in a more useful form. We take the natural log of both sides and apply Stirling's approximation, $\ln N! = N \ln N - N$, to get

$$\ln P(s) \approx N \ln 2 - \frac{1}{2}(N+s)\ln(1+\frac{s}{N}) - \frac{1}{2}(N-s)\ln(1-\frac{s}{N}). \tag{13.25}$$

We know that $P(s)$ will be sharply peaked at $s = 0$ for large N from its definition, equation (13.24), and from common sense. Using the expansion

$$\ln(1+x) = x - \frac{1}{2}x^2 + \cdots, \tag{13.26}$$

we have

$$\ln P(s) \approx N \ln 2 - \frac{s^2}{2N} + \mathcal{O}\left(\frac{s^4}{N^3}\right), \tag{13.27}$$

or

$$P(s) = 2^N e^{-s^2/2N}. \tag{13.28}$$

Changing variables from s to $x = as$ gives

$$P(x) = \frac{1}{a\sqrt{2N\pi}}e^{-x^2/2Na^2}, \tag{13.29}$$

where we have normalized $P(x)$ such that

$$1 = \int_{-\infty}^{+\infty} P(x)dx. \tag{13.30}$$

As we expect for a random walk, the probability distribution is a gaussian centered at $x = 0$ with a standard deviation of $\sqrt{N}a$.

b) If we eliminate the number of jumps N by writing $N = t/\tau$, the probability distribution becomes

$$P(x,t) = \sqrt{\frac{\tau}{2\pi a^2 t}}e^{-x^2\tau/2a^2 t}. \tag{13.31}$$

Next, we wish to establish the connection between $P(x,t)$ and the concentration of impurities in the lattice, $f(x,t)$. Imagine M impurities in the lattice at arbitrary initial positions x_0^1, x_0^2,...,x_0^M, at time $t = 0$. Then

$$f(x,t) = \sum_{i=1}^{M} P(x - x_0^i, t) \qquad (13.32)$$

is the probability distribution for finding a particle at x at time t. For large N, $f(x)$ is the actual concentration of impurities. Since $f(x,t)$ is just a sum of probabilities of the form (13.29), to find D it is sufficient to substitute expression (13.29) for $P(x,t)$, rather than $f(x,t)$, into

$$D\frac{\partial^2 f}{\partial x^2} = \frac{\partial f}{\partial t}. \qquad (13.33)$$

Doing this, we find that equation (13.33) is satisfied if $D = a^2/2\tau$.

Note that although $P(x,t)$ was constructed without regard to the differential equation (13.33), it nevertheless satisfies this equation. (An arbitrary function $g(x,t)$ would clearly not satisfy this equation for constant D). We were not just lucky. Both the differential equation (13.33) and $P(x,t)$ were constructed under the assumptions of conserved particle number and random diffusion. In deriving $P(x,t)$, the assumption of random diffusion was implicit in equation (13.23), and particle number conservation was ensured by normalizing $P(x,t)$. To show the role of these assumptions in the construction of equation (13.33) for the diffusivity D, we sketch its derivation below.

Impurity number conservation implies the existence of a continuity equation,

$$\frac{\partial f}{\partial t} + \boldsymbol{\nabla} \cdot \mathbf{j} = 0, \qquad (13.34)$$

where $\mathbf{j}$ is the flux of impurity atoms. If the flux of impurity atoms is due solely to a gradient in the concentration (i.e., random diffusion), then

$$\mathbf{j} = -D\boldsymbol{\nabla} f(x,t). \qquad (13.35)$$

(Normally one might add terms on the right-hand side from the effects of other driving forces, e.g., an electric field). Inserting this expression for $\mathbf{j}$ into the continuity equation gives equation (13.33).

Solution 4.4. The Clapeyron equation, which describes the phase coexistence curve, is

$$\frac{dP}{dT} = \frac{L}{T\Delta v},$$

(13.36)

where $P(T)$ gives the phase boundary in P-T space, and

$$v_l - v_s \equiv \Delta v = \frac{1}{\rho_l} - \frac{1}{\rho_s} = \frac{\rho_s - \rho_l}{\rho_l \rho_s} \equiv \frac{\Delta\rho}{\rho_l \rho_s}$$

(13.37)

is the difference in the volume per unit mass in the two phases. Note that $L = T(S_l - S_s)$, the amount of heat required to melt a unit of mass, is a positive quantity.

Let us consider the pressure to be a function of h, rather than temperature. We write

$$\frac{dP}{dT} = \frac{dP}{dh}\frac{dh}{dT},$$

(13.38)

so that the Clapeyron equation (13.36) becomes

$$\left(\frac{\Delta\rho}{\rho_l \rho_s}\right)\frac{dP}{dh} = \frac{L}{T(dh/dT)}.$$

(13.39)

Consider the derivative dP/dh. The differential dP is the change in the pressure at the interface if the interface is raised a height dh. Since the total mass in the column is conserved, if h increases by dh, there will be $A\rho_s dh$ less mass pushing down on the interface (where A is the cross-sectional area of the column), so that we may write

$$\frac{dP}{dh} = -\rho_s g.$$

(13.40)

We use this with equation (13.39) to find

$$\Delta\rho = \frac{-L\rho_l}{gT(dh/dT)}.$$

(13.41)

Solution 4.5. From the standard thermodynamic relation

$$dU = T\,dS - P\,dV, \tag{13.42}$$

we note that the work done by the engine during one cycle is

$$
\begin{aligned}
W_{tot} &= W_{A \to B} + W_{C \to D} \\
&= \int_{A \to B} P\,dV + \int_{C \to D} P\,dV,
\end{aligned} \tag{13.43}
$$

and the energy absorbed by the engine is

$$
\begin{aligned}
Q_{in} &= Q_{B \to C} \\
&= \int_{B \to C} T\,dS.
\end{aligned} \tag{13.44}
$$

(The heat lost by the engine between D and A is wasted energy.) The efficiency we wish to find is defined in terms of these quantities by

$$\epsilon \equiv \frac{W_{tot}}{Q_{in}}. \tag{13.45}$$

Using the two ideal gas relations $PV = NkT$ and $C_V\,dT = dU$, equation (13.42) can be recast as

$$dS = C_V \frac{dT}{T} + Nk \frac{dV}{V}, \tag{13.46}$$

where k is the Boltzmann constant. We integrate this to find the entropy:

$$S = C_V \ln T + Nk \ln V + \text{constant}. \tag{13.47}$$

Solving this for the temperature, we find

$$T = \alpha e^{S/C_V} V^{-Nk/C_V}, \tag{13.48}$$

where α is a constant with the appropriate dimensions. We can now find the work done by the system between A and B. Since we are compressing the gas, we are actually doing work *on* the system and $W_{A \to B}$ is negative:

$$
\begin{aligned}
W_{A \to B} &= \int_{V_A}^{V_B} P \, dV \\
&= -Nk \int_{V_B}^{V_A} \frac{T}{V} \, dV \\
&= -\alpha Nk \int_{V_B}^{V_A} e^{S_2/C_V} V^{-(1+Nk/C_V)} \, dV \\
&= -\alpha \left(V_B^{-Nk/C_V} - V_A^{-Nk/C_V} \right) e^{S_2/C_V} C_V. \quad (13.49)
\end{aligned}
$$

We get a similar expression (with opposite sign) for $W_{C \to D}$. Putting these together, we find that the total work done in one cycle is

$$
W_{tot} = \alpha C_V \left(V_B^{-Nk/C_V} - V_A^{-Nk/C_V} \right) \left(e^{S_1/C_V} - e^{S_2/C_V} \right). \quad (13.50)
$$

The heat of combustion can be found similarly:

$$
\begin{aligned}
Q_{B \to C} &= \int_{S_2}^{S_1} T \, dS \\
&= \alpha \int_{S_2}^{S_1} V_B^{-Nk/C_V} e^{S/C_V} \, dS \\
&= \alpha C_V V_B^{-Nk/C_V} \left(e^{S_1/C_V} - e^{S_2/C_V} \right). \quad (13.51)
\end{aligned}
$$

From equation (13.45) we can now find the efficiency of the engine,

$$
\epsilon = 1 - \left(\frac{V_A}{V_B} \right)^{-k/c_V}, \quad (13.52)
$$

where we have written $c_V = C_V/N$. We note that in the special case of a monatomic ideal gas, for which $c_V = 3k/2$, we have

$$
\epsilon = 1 - \left(\frac{V_A}{V_B} \right)^{-2/3}. \quad (13.53)
$$

Solution 4.6. First we will carefully derive a rigorous result; however, at the end of the solution we also include a simple method for deriving the leading force term to within a constant factor.

For a gas in thermal equilibrium at temperature T, the number of molecules per unit volume with velocity $\mathbf{v}$ is given by the Maxwellian distribution,

$$\rho(\mathbf{v}) = n \left(\frac{m}{2\pi kT} \right)^{3/2} \exp \left(-\frac{m\mathbf{v}^2}{2kT} \right), \qquad (13.54)$$

where n is the number of molecules per unit volume and m is the mass of one molecule. Suppose the disk has velocity $\mathbf{V} = v_0 \hat{\mathbf{x}}$, with the disk lying in the yz-plane. It is evident that we need consider only the motions of the molecules in the x-direction. Therefore we do not need the full distribution (13.54), but rather only the distribution of v_x,

$$\tilde{\rho}(v_x) = \int \int dv_y \, dv_z \, \rho(\mathbf{v}) = n \left(\frac{m}{2\pi kT} \right)^{1/2} \exp \left(-\frac{mv_x^2}{2kT} \right). \qquad (13.55)$$

Consider a molecule with velocity v_x in the region behind the disk. It will catch up and collide with the disk if its velocity relative to the disk, $v_{rel} = v_x - v_0$, is positive. Because the collision is elastic, the new relative velocity of the molecule will be $\hat{v}_{rel} = -v_{rel}$. If we assume that the mass of the molecule is far less than that of the disk, the molecule's change in velocity will be $\Delta v = -2v_{rel} = -2(v_x - v_0)$. (Note that after colliding with the disk, the molecules are no longer in thermal equilibrium with the rest of the gas. For this reason our eventual answer will depend on the assumption that the mean free path of the molecules is much greater than the size of the disk, so that the molecules hitting the disk had their last collision in a region that *was* in thermal equilibrium.)

The impulse imparted to the disk in this collision is $2m(v_x - v_0)$. The number of molecules with velocity v_x that will collide with the area πR^2 in unit time is

$$N(v_x) = \pi R^2 \tilde{\rho}(v_x)(v_x - v_0). \qquad (13.56)$$

Therefore the total impulse per unit time due to molecules colliding with the back of the disk is

$$I_1 = \int_{v_0}^{\infty} dv_x \, 2\pi m R^2 \, \tilde{\rho}(v_x)(v_x - v_0)^2. \qquad (13.57)$$

Now consider the area in front of the disk. A molecule in this region will collide with the disk if it has velocity $v_x < v_0$, and the impulse it imparts to the disk will be $-2m(v_0 - v_x)$. The total impulse per unit time on the front of the disk is

$$I_2 = - \int_{-\infty}^{v_0} dv_x \, 2\pi m R^2 \, \tilde{\rho}(v_x)(v_0 - v_x)^2. \qquad (13.58)$$

Consequently the net impulse per unit time, i.e., the net force, will be

$$F = I_1 + I_2 = -2\pi m R^2 \left\{ \int_{-\infty}^{v_0} dv_x \, \tilde{\rho}(v_x)(v_0 - v_x)^2 - \right.$$
$$\left. \int_{v_0}^{\infty} dv_x \, \tilde{\rho}(v_x)(v_0 - v_x)^2 \right\}. \quad (13.59)$$

It will simplify matters if we substitute $u = -v_x$ in the first integral (and $u = v_x$ in the second), and note that $\tilde{\rho}(u) = \tilde{\rho}(-u)$. Then the first integral becomes

$$\int_{-\infty}^{v_0} dv_x \, \tilde{\rho}(v_x)(v_0 - v_x)^2 = \int_{-v_0}^{\infty} du \, \tilde{\rho}(u)(v_0 + u)^2. \qquad (13.60)$$

We can now rearrange (13.59) into a more manageable form:

$$F = -2\pi m R^2 \left\{ \int_{-v_0}^{v_0} du \, \tilde{\rho}(u)(v_0 + u)^2 + \right.$$
$$\left. \int_{v_0}^{\infty} du \, \tilde{\rho}(u) \left[(v_0 + u)^2 - (v_0 - u)^2 \right] \right\}. \quad (13.61)$$

To evaluate the first integral, we can use the assumption that the speed v_0 of the disk is far smaller than the average molecular speed $\bar{v}$. The average speed of a gas molecule is given by the equipartition theorem as $m\bar{v}^2 \approx kT$. This allows us to approximate the Boltzmann factor in $\tilde{\rho}(u)$ by unity, because

$$\exp\left(-\frac{mv_0^2}{2kT}\right) \approx \exp\left(-\frac{v_0^2}{\bar{v}^2}\right) \approx 1. \qquad (13.62)$$

Hence we can set $\tilde{\rho}(u)$ to $\tilde{\rho}(0) = n(m/2\pi kT)^{1/2}$ within the range of integration. The second integral can be evaluated explicitly, because $(v_0 + u)^2 - (v_0 - u)^2 = 4v_0 u$. Therefore

$$F = -2\pi m R^2 n \left(\frac{m}{2\pi kT}\right)^{1/2} \left\{\int_{-v_0}^{v_0} du\,(v_0 + u)^2 + \right.$$

$$\left. \int_{v_0}^{\infty} du\,4v_0 u \exp\left(-\frac{mu^2}{2kT}\right)\right\}$$

$$= -2\pi m R^2 n \left(\frac{m}{2\pi kT}\right)^{1/2} \left\{\frac{8}{3}v_0^3 + \frac{4kT}{m}v_0 \exp\left(-\frac{mv_0^2}{2kT}\right)\right\}. \quad (13.63)$$

Now, by previous assumption, $mv_0^2 \ll kT$, so we can expand the exponential and get our final result:

$$F = -2\pi m R^2 n \left(\frac{m}{2\pi kT}\right)^{1/2} \left\{\frac{4kT}{m}v_0\left(1 - \frac{mv_0^2}{2kT} + \cdots\right) + \frac{8}{3}v_0^3\right\}$$

$$= -(2\pi kTm)^{1/2}4nR^2 v_0 \left\{1 + \frac{mv_0^2}{6kT} + \cdots\right\}. \quad (13.64)$$

As promised, we now describe a simple method of finding the leading term in the force to within a numerical factor. We make the approximation that all molecules are traveling with velocity $\bar{v}$. After colliding with the disk, the molecules on the front side have speed $\bar{v} + v_0$, and those on the back side have speed $\bar{v} - v_0$. So the average molecule gains in kinetic energy by an amount

$$\Delta KE = \frac{1}{2}m\left(\frac{1}{2}\left[(\bar{v} + v_0)^2 + (\bar{v} - v_0)^2\right] - \bar{v}^2\right) = \frac{1}{2}mv_0^2. \quad (13.65)$$

Since $v_0 \ll \bar{v}$, the disk collides with $n\pi R^2 \bar{v}t$ molecules in time t; in this time it will lose a kinetic energy of $\frac{1}{2}mv_0^2 \times n\pi R^2 \bar{v}t$ to the gas. This is the work done by the drag force, so

$$F v_0 t = \frac{1}{2}mv_0^2 \times n\pi R^2 \bar{v}t. \quad (13.66)$$

If we substitute $\bar{v}^2 = 3kT/m$, we find the drag is

$$F = \frac{\pi}{2}\sqrt{3kTmn}R^2 v_0, \quad (13.67)$$

which is close to our original result.

Solution 4.7.   First, we need to find the equation of motion, and its solutions $y(x, t)$, for the symmetric normal modes of the wire. (The antisymmetric modes do not move the midpoint of the wire.) The force on an element Δx of the wire is

$$F = \tau \frac{dy}{dx}\bigg|_{x+\Delta x} - \tau \frac{dy}{dx}\bigg|_{x} = \tau \frac{d^2y}{dx^2}\Delta x. \tag{13.68}$$

We use this result and Newton's law,

$$F = (\mu \Delta x)\frac{d^2y}{dt^2}, \tag{13.69}$$

to find the equation of motion:

$$\frac{d^2y}{dt^2} = \frac{\tau}{\mu}\frac{d^2y}{dx^2}. \tag{13.70}$$

This equation has solutions $y(x, t) = A_n \sin(\omega_n t + \alpha) \sin k_n x$, where

$$\frac{\omega_n^2}{k_n^2} = \frac{\tau}{\mu}, \tag{13.71}$$

and where $k_n = n\pi/l$ so that $y(0) = y(l) = 0$. We are only interested in symmetric modes, where n is odd. The kinetic energy in each mode is

$$\epsilon_n = \int_0^l \frac{1}{2}\mu \left(\frac{dy}{dt}\right)^2 dx. \tag{13.72}$$

Since the energy in each mode is given by the maximum kinetic energy in the mode, we want to evaluate ϵ_n at a time t such that dy/dt is a maximum, or where $\cos(\omega_n t + \alpha) = 1$. If we also recall that the integral $\int_0^{n\pi} \sin^2 w \, dw$ is equal to $n\pi/2$, and use the relation between ω_n and k_n, (13.71), we find that the energy in the mode is

$$E_n = \frac{\tau n^2 \pi^2}{4l} A_n^2. \tag{13.73}$$

Each mode contributes two degrees of freedom (potential energy and kinetic energy), so that in thermal equilibrium

$$E_n = 2(\frac{1}{2}k_BT) = k_BT. \tag{13.74}$$

From the last two equations, we may solve for A_n^2, leading to

$$A_n^2 = \frac{4lk_BT}{\tau\pi^2n^2}. \tag{13.75}$$

The square of the fluctuations of the midpoint of the wire is given by:

$$\langle y_{mid}^2 \rangle = \left\langle \left(\sum_{n \text{ odd}} A_n \sin(\omega_n t + \alpha) \sin(n\pi/2) \right)^2 \right\rangle. \tag{13.76}$$

Here, the angle brackets denote a time average. Notice that when we average over time the cross terms will vanish. Thus we have

$$\langle y_{mid}^2 \rangle = \left[\frac{1}{2} \sum_{n \text{ odd}} A_n^2 \right] = \left[\frac{4lk_BT}{2\tau\pi^2} \sum_{n \text{ odd}} \frac{1}{n^2} \right]. \tag{13.77}$$

Using the given sum, namely

$$\sum_{m=0}^{\infty} (2m+1)^{-2} = \frac{\pi^2}{8}, \tag{13.78}$$

and taking the square root, we find the rms fluctuation:

$$y_{rms} = \frac{1}{2}\sqrt{\frac{lk_BT}{\tau}}. \tag{13.79}$$

If we had wanted to answer the question quickly, we could have estimated y_{rms} using only the contribution from the fundamental ($n = 1$) mode, which is dominant.

Solution 4.8.   Our approach will be to find expressions for the chemical potentials of the gas μ_g and of the adsorbed film on the surface μ_s, and equate them in order to solve for the vapor pressure P_g. The chemical potential μ is defined in general by

$$\mu \equiv \frac{\partial F(N,T)}{\partial N}, \tag{13.80}$$

where the free energy F is given by

$$F = -k_B T \ln Z, \tag{13.81}$$

with Z the partition function.

First, we find the partition function for one particle of the vapor:

$$Z_1 = \frac{1}{h^3} \int e^{-\mathbf{p}^2/2mk_B T} \, d^3p \, d^3x = \frac{V}{h^3} \left[\int e^{-p_x^2/2mk_B T} \, dp_x \right]^3. \tag{13.82}$$

This results in

$$Z_1 = \frac{V}{h^3} (2\pi m k_B T)^{\frac{3}{2}}. \tag{13.83}$$

The *total* partition function for the vapor is

$$Z_g = \frac{1}{N_g!} Z_1^{N_g}, \tag{13.84}$$

where N_g is the number of particles in the vapor. From equations (13.80) and (13.81), and using Stirling's approximation, we find

$$\mu_g \approx -k_B T \ln \left\{ \frac{V_g}{N_g} \left(\frac{mk_B T}{2\pi\hbar^2} \right)^{\frac{3}{2}} \right\}. \tag{13.85}$$

We then use the equation of state for an ideal gas, $P_g V_g = N_g k_B T$, to write

$$\mu_g \approx -k_B T \ln \left\{ \frac{k_B T}{P_g} \left(\frac{mk_B T}{2\pi\hbar^2} \right)^{\frac{3}{2}} \right\}. \tag{13.86}$$

Now we need an expression for μ_s. Consider n particles distributed on N sites. The number of possible states is

$$g = \frac{N!}{(N-n)!\,n!}. \tag{13.87}$$

The partition function is given by the sum of the Boltzmann factors, which are identical for a given n:

$$Z(n) = \frac{N!}{(N-n)!\,n!} e^{-n(-\epsilon_0)/k_B T}. \tag{13.88}$$

We must be careful about signs here. The atoms bind to the surface with an energy $-\epsilon_0$, and thus the exponential in the partition function is $e^{+n\epsilon_0/k_B T}$.

Finding an expression for $\langle n \rangle$ will give us a relationship between μ_s and n/N. Since we will be considering a system that does not have a fixed number of particles, we use the grand canonical ensemble, where

$$\langle n \rangle = k_B T \frac{\partial}{\partial \mu_s} (\ln \Xi) = \frac{1}{\Xi} k_B T \frac{\partial \Xi}{\partial \mu_s}. \tag{13.89}$$

Here, Ξ is the grand canonical partition function,

$$\Xi = \sum_{n=0}^{n=N} \lambda^n Z(n) = (1 + e^{(\mu_s + \epsilon_0)/k_B T})^N, \tag{13.90}$$

where $\lambda = e^{\mu_s/k_B T}$, and we have used $Z(n)$ from equation (13.88). This leads to

$$\frac{\langle n \rangle}{N} = \frac{1}{1 + e^{-(\mu_s + \epsilon_0)/k_B T}}, \tag{13.91}$$

or

$$\mu_s = -k_B T \ln \left\{ \left(\frac{N}{\langle n \rangle} - 1 \right) e^{\epsilon_0/k_B T} \right\}. \tag{13.92}$$

Upon equating μ_s and μ_y, and taking $n = \langle n \rangle$, we find

$$P_g = \left(\frac{mk_B T}{2\pi \hbar^2} \right)^{3/2} \frac{k_B T e^{-\epsilon_0/k_B T}}{N/n - 1}. \tag{13.93}$$

Solution 4.9. a) In a system with molar entropy, volume, and internal energy s, v, and u, the first law of thermodynamics has the form $dU = T\,ds - P\,du$. The Clapeyron equation gives the coexistence curve $P_c(T)$ in terms of the changes Δs and Δv between the two phases:

$$\frac{dP_c}{dT} = \frac{\Delta s}{\Delta v}. \tag{13.94}$$

Similarly, a material in a magnetic field has $du = T\,ds + H\,dm$, where m is the magnetic moment per mole. Thus the Clapeyron equation analogous to (13.94) for a type I superconductor is

$$\frac{dH_c}{dT} = \frac{s_n - s_s}{m_s - m_n} = -\frac{\Delta s}{\Delta m}. \tag{13.95}$$

Since from the third law of thermodynamics $s \to 0$ on both sides of the phase boundary as $T \to 0$, Δs must vanish as well. Thus, from (13.95) we see

$$\frac{dH_c}{dT} \to 0 \quad \text{as} \quad T \to 0, \tag{13.96}$$

which implies that $a = 0$.

b) The Clapeyron equation can be written

$$\frac{dH_c}{dT} = \frac{-l}{T\Delta m}, \tag{13.97}$$

where $l \equiv T\Delta s$ is the latent heat per mole. Previously, we considered l and m to be molar quantities. We now redefine them to be quantities per unit volume (without changing the equation). Taking the derivative (and remembering that $a = 0$), we find that

$$l = -2bT^2\Delta m, \tag{13.98}$$

where $\Delta m \equiv m_s - m_n$.

We have to find b. At $T = T_c$,

$$H_c = H_0 + bT_c^2 = 0, \tag{13.99}$$

which gives $b = -H_0/T_c^2$. Substituting b into (13.98) yields

$$l = 2H_0 \left(\frac{T}{T_c}\right)^2 \Delta m = 2H_0 \left(\frac{T}{T_c}\right)^2 m_s, \qquad (13.100)$$

where m_s is the magnetization inside the superconductor, and we have set the magnetization in the normal state equal to zero.

Next we must find m_s. Inside a type I superconductor $\mathbf{B} = \mathbf{H} + 4\pi\mathbf{m} = 0$, so that $m_s = -H_c/4\pi$. We use the parabolic form (13.99) for H_c and our latest expression for the latent heat, (13.100), to arrive at the answer:

$$l = -\frac{H_0^2}{2\pi} \left[\left(\frac{T}{T_c}\right)^2 - \left(\frac{T}{T_c}\right)^4 \right], \qquad (13.101)$$

where l is the latent heat going from the normal to the superconducting state.

c) We relate the discontinuity in the specific heat per unit volume,

$$\Delta c_H = T \left[\left(\frac{\partial s_s}{\partial T}\right)_H - \left(\frac{\partial s_n}{\partial T}\right)_H \right] = T \left(\frac{\partial \Delta s}{\partial T}\right)_H, \qquad (13.102)$$

to the latent heat (whose temperature dependence we know) by using the definition $l \equiv T\Delta s$:

$$\Delta c_H = T \frac{\partial(l/T)}{\partial T}. \qquad (13.103)$$

Taking the derivative leads to

$$\Delta c_H = -\frac{H_0^2}{2\pi} \left[\frac{T}{T_c^2} - \frac{3T^3}{T_c^4} \right]. \qquad (13.104)$$

Solution 4.10. The entropy S of a state is defined as the natural logarithm of the degeneracy of the state, or $S = k \ln g$, where k is Boltzmann's constant and g is the degeneracy.

a) The number of different ways of picking the n atoms which jump to the interstitial sites is given by

$$\binom{N}{n} \equiv \frac{N!}{n!(N-n)!}. \tag{13.105}$$

Each one of these atoms could be in one of eight positions, yielding a total degeneracy of

$$g = 8^n \binom{N}{n}. \tag{13.106}$$

Here we have assumed that since $n \ll N$, we never encounter two atoms attempting to jump to the same intersitial site. The entropy is therefore

$$S = k \left[n \ln 8 + \ln \left(\frac{N!}{n!(N-n)!} \right) \right]. \tag{13.107}$$

Using Stirling's approximation for $\ln n!$, and dropping terms of order (n/N), we find:

$$S \approx kn(1 + \ln 8 + \ln \frac{N}{n}) \approx kn \ln \frac{N}{n}. \tag{13.108}$$

b) Assume there is no correlation between the unoccupied O-sites and the occupied X-sites. Then the total degeneracy of the system is the product of the degeneracy of the O-lattice of holes with the degeneracy of the X-lattice of occupied sites, each of which is given by equation (13.105), so that

$$g = \binom{N}{n}^2. \tag{13.109}$$

Following the same procedure as above we find

$$S \approx 2kn \left(1 + \ln \frac{N}{n} \right) \approx 2kn \ln \frac{N}{n}. \tag{13.110}$$

c) We will describe two approaches to this problem. In the thermodynamic approach, we recall one definition of temperature:

$$\left(\frac{\partial S}{\partial U} \right)_V = \frac{1}{T}. \tag{13.111}$$

Since $U = n\epsilon$, this can be rewritten as

$$\frac{\partial S}{\partial n} = \frac{\epsilon}{T}.$$ (13.112)

For case (a), we use the expression for entropy (13.108) in the above equation to find the fraction of occupied interstitial sites:

$$\frac{n}{N} = 8e^{-\epsilon/kT}.$$ (13.113)

For case (b), we use the entropy given by equation (13.110) to find:

$$\frac{n}{N} = e^{-\epsilon/2kT}.$$ (13.114)

Of course, we can also obtain these answers using statistical mechanics. The probability that a system is in a state with degeneracy g and energy E is

$$P = \frac{ge^{-E/kT}}{Z},$$ (13.115)

where Z is the (constant) partition function. Therefore for case (a) we can write the probability that the system is in a state with n atoms excited as

$$P(n) \propto \binom{N}{n} 8^n e^{-\beta n\epsilon}.$$ (13.116)

This probability will be sharply peaked at one particular value of n, $\bar{n}$, which will be the observed value of n. Since $P(n)$ and $\ln P(n)$ each have a maximum at $\bar{n}$, it is easiest to use the condition

$$\frac{\partial}{\partial n} \left[\ln P(n)\right]_{n=\bar{n}} = 0$$ (13.117)

to find $\bar{n}$, because one can then make use of Stirling's approximation. Again we are led to the fraction of occupied interstitial sites given in equation (13.113).

For case (b), the probability of the system having n occupied interstitial sites is

$$P(n) \propto \binom{N}{n}^2 e^{-\beta n\epsilon}.$$ (13.118)

Again, we recover our previous result.

Chapter 14

Condensed Matter Physics—Solutions

Solution 5.1. This problem requires the use of the "semiconductor model" for superconductors. For further discussion of this model, see Tinkham, Chapter 2.

a) The goal is to find an expression for the current I in terms of the voltage V. To get this expression, we will use Fermi's golden rule to find the rate of tunneling of electrons from metal 1 to metal 2, and vice versa. The net current is due to the difference between these tunneling rates.

Let us write the chemical potential of electrons in metal 1 as μ_1 and in metal 2 as μ_2. Since metal 2 is at a voltage $+V$ and metal 1 is grounded, these chemical potentials satisfy $\mu_1 = \mu_2 + eV$. (Don't forget that electrons are negatively charged, so they will flow from metal 1 to metal 2.)

The occupation probability P for an electron of energy ϵ in metal i is given by

$$P(\epsilon) = f(\epsilon - \mu_i) = \frac{1}{e^{(\epsilon - \mu_i)/kT} + 1}, \tag{14.1}$$

where $f(\epsilon)$ is the Fermi function. The density of states at energy ϵ in metal i is $n_i(\epsilon - \mu_i)$. Let us introduce the variable $E = \epsilon - \mu_1$, which

166

is the electron energy (in either metal) measured with respect to the chemical potential of metal 1. Consider an electron in metal 2 with energy E. The rate of tunneling $R(E)$ into metal 1 for this electron is given by Fermi's golden rule as

$$R(E) \propto |M|^2 \rho(E), \qquad (14.2)$$

where $\rho(E)$ is the density of *available* final states (in metal 1) at energy E, and M is the given tunneling amplitude. The value of $\rho(E)$ is the density of states multiplied by the probability that the states are unoccupied,

$$[1 - f(\epsilon - \mu_1)]n_1(\epsilon - \mu_1) = [1 - f(E)]n_1(E). \qquad (14.3)$$

Thus the rate of tunneling per electron in metal 2, as a function of the electron's energy E, is

$$R(E) \propto |M|^2[1 - f(E)]n_1(E). \qquad (14.4)$$

The total number of electrons at energy E in metal 2 that might tunnel is $n_2(E + eV)f(E + eV)$. Integrating over all possible energies E, we find the current from 2 to 1:

$$I_{2\to1} = -eK \int_{-\infty}^{\infty} |M|^2 n_2(E+eV)f(E+eV)n_1(E)[1-f(E)]\, dE, \qquad (14.5)$$

where K is a constant that depends on the geometry of the junction. To find the net current, we also need the current $I_{1\to2}$ from metal 1 to metal 2. Using the same arguments as above, we find

$$I_{1\to2} = -eK \int_{-\infty}^{\infty} |M|^2 n_1(E)f(E)n_2(E+eV)[1-f(E+eV)]\, dE. \qquad (14.6)$$

The net current between the two metals is $I = I_{2\to1} - I_{1\to2}$, so

$$I = eK \int_{-\infty}^{\infty} |M|^2 n_1(E)n_2(E + eV)[f(E) - f(E + eV)]\, dE. \qquad (14.7)$$

Notice that this result is valid for general densities of states $n(E)$.

Consider the factor $[f(E) - f(E + eV)]$. At low temperatures, this has the limiting form shown in Figure 14.1. Because eV is much smaller

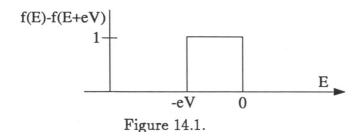

Figure 14.1.

than $E_F \approx \mu_i$, n_1 and n_2 do not change very much in the region in which this factor (and thus the integrand) is nonzero. Therefore, we can evaluate each density-of-states function at the Fermi energy of the metal, corresponding to $\epsilon = \mu_i$, and bring them both outside the integral. We may rewrite $[f(E) - f(E + eV)] \approx -eV(df/dE)$. Recall that

$$\int_{-\infty}^{\infty} \left(-\frac{df}{dE} \right) dE = f(-\infty) - f(\infty) = 1, \qquad (14.8)$$

regardless of the temperature T, because of the form of $f(E)$. (Note that this can also be written down for $T = 0$ by inspection of Figure 14.1.) Thus we find $I = K|M|^2 n_1(0) n_2(0) e^2 V$, which gives the conductance G:

$$G = e^2 K|M|^2 n_1(0) n_2(0). \qquad (14.9)$$

b) We emphasize that we are using the *semiconductor model*, in which at $T = 0$ the superconductor has a bandgap of 2Δ with no occupied states above the bandgap, and a full sea of occupied states below the bandgap. See, for example, Rose-Innes and Rhoderick for a discussion of this model.

We will again use equation (14.7) for the net current, although with a different density-of-states function to describe the superconductor. As in part (a), we evaluate the density of states of the normal metal at the Fermi energy and bring it outside the integral:

$$I = eK'|M|^2 n_2(0) \int_{-\infty}^{\infty} g(E)[f(E) - f(E + eV)] \, dE. \qquad (14.10)$$

Here, g denotes the density of single-particle states of the superconductor, metal 1. We reserve n for densities of states of normal metals.

Since we are considering this current at zero temperature, the form of $[f(E) - f(E + eV)]$ is exactly as pictured in Figure 14.1. Thus, we may rewrite the current as

$$I = eK'|M|^2 n_2(0) \int_{-eV}^{0} g(E) \, dE. \qquad (14.11)$$

We must now borrow a result from BCS theory giving the relationship between the density of states for quasiparticles in a superconductor and the density of states for electrons in a normal metal:

$$\frac{g(E)}{n_1(0)} = \begin{cases} \dfrac{|E|}{(E^2 - \Delta^2)^{1/2}} & \text{if } |E| > \Delta, \text{ or} \\ 0 & \text{otherwise.} \end{cases} \qquad (14.12)$$

Thus we may rewrite the integral as

$$\int_{-eV}^{0} g(E) dE = \begin{cases} n_1(0) \displaystyle\int_{-eV}^{-\Delta} \dfrac{|E| dE}{(E^2 - \Delta^2)^{1/2}} & \text{if } eV > \Delta, \text{ or} \\ 0 & \text{otherwise,} \end{cases} \qquad (14.13)$$

which gives us the desired current:

$$I = \begin{cases} eK'|M|^2 n_2(0) n_1(0)(e^2 V^2 - \Delta^2)^{1/2} & \text{if } eV > \Delta, \text{ or} \\ 0 & \text{otherwise.} \end{cases} \qquad (14.14)$$

Note that unlike the conductivity for a metal-metal junction (14.9), which only tells us about the factor K, the conductivity for a metal-superconductor junction contains information about the superconductor, through its dependence on the energy gap Δ.

c) This part of the problem is most easily understood through pictures (Figure 14.2). We assume that T is small but finite, so that both metals are in their superconducting states and have a small number of states occupied above their bandgaps. At $V = 0$ (see Figure 14.2a), there is virtually no tunneling between the two metals because there are very few quasiparticles above the bandgap (and, similarly, very few quasiparticle holes below the bandgap). It is only quasiparticles *above* (or holes below) the bandgap which can tunnel into unoccupied states at the same energy in the neighboring superconductor. When the voltage reaches $eV_1 = \Delta_{Pb} - \Delta_{Al}$ (see Figure 14.2b) the edges of the bands

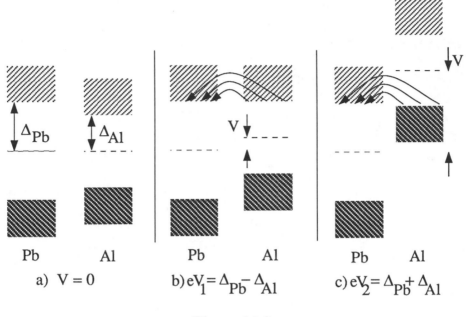

Figure 14.2.

of the two metals are at the same energy. Recall that the density-of-states functions (14.12) blow up at the band edge. Therefore most of the excited quasiparticles are very near the band edge and, again because the density of states is large in the neighboring superconductor, the number of available states into which these quasiparticles can tunnel is enormous. Therefore there is a local maximum in the current at V_1. If the voltage continues to increase, eventually the lower band edge of the aluminum bandgap reaches the upper band edge of the lead bandgap, Figure 14.2c, and the current rises very rapidly because there are now a huge number of occupied states in aluminum that can tunnel. This happens when $eV_2 = \Delta_{Pb} + \Delta_{Al}$. We now have two equations in two unknowns, so we may solve for both bandgaps:

$$\Delta_{Pb} = \frac{e}{2}(V_2 + V_1) = 13.5 \times 10^{-4} \text{ eV} \qquad (14.15)$$

$$\Delta_{Al} = \frac{e}{2}(V_2 - V_1) = 1.7 \times 10^{-4} \text{ eV}. \qquad (14.16)$$

Solution 5.2. At the onset of superconductivity, we expect that the number of Cooper pairs $n_s = \psi^* \psi$ will be small, so we will neglect the last term in the Landau-Ginzburg equation, which is cubic in ψ. The magnetic field $\mathbf{B}$ in the sample will be dominated by the external magnetic field $\mathbf{B}_a$, so $\mathbf{B} \approx \mathbf{B}_a$. Choose the gauge in which $\mathbf{A} = B(0, x, 0)$. Equation (5.1) then becomes

$$\left(\frac{\partial^2}{\partial x^2} + \frac{\partial^2}{\partial z^2} \right) \psi + \left(\frac{\partial}{\partial y} - i \frac{2eBx}{\hbar c} \right)^2 \psi = -\xi^{-2} \psi, \qquad (14.17)$$

which is reminiscent of the Schrödinger equation for a free particle in a magnetic field (see Problem 5.10).

In order to solve this equation, we make the substitution $\psi(\mathbf{x}) = \phi(x) \exp(ik_y y) \exp(ik_z z)$. After some algebra and the further definition $\tilde{x} = x - \hbar c k_y / 2eB$, we have

$$\left(-\frac{\hbar^2}{2} \frac{d^2}{d\tilde{x}^2} + \frac{2e^2 B^2}{c^2} \tilde{x}^2 \right) \phi(\tilde{x}) = \left(\frac{\hbar^2}{2\xi^2} - \frac{\hbar^2 k_z^2}{2} \right) \phi(\tilde{x}), \qquad (14.18)$$

which is formally the equation of a simple harmonic oscillator of unit mass:

$$\left(-\frac{\hbar^2}{2} \frac{d^2}{dx^2} + \frac{1}{2} \omega^2 x^2 \right) \phi = E\phi, \qquad (14.19)$$

where $\omega = 2eB/c$ is the angular frequency. We know all about the solutions of this equation. In particular the allowed energies are $E_n = (n + 1/2)\hbar\omega$, so we can immediately equate

$$\frac{\hbar^2}{2\xi^2} - \frac{\hbar^2 k_z^2}{2} = (n + 1/2) \frac{2\hbar eB}{c}. \qquad (14.20)$$

For a given ξ, we see that there is a maximum allowed value of the magnetic field, B_{max}, corresponding to $n = 0$ and $k_z = 0$. For fields greater than this maximum, the sample is no longer superconducting. The value of the maximum field is

$$H_{c2} \equiv B_{max} = \left(\frac{\hbar c}{2e} \right) \frac{1}{\xi^2} = \frac{\hbar c}{2e\xi_0^2} \left(1 - \frac{T}{T_c} \right). \qquad (14.21)$$

Note that we could also have estimated B_{max} by requiring that it be the field at which vortices in the superconductor intersect. The radius of a vortex is approximately ξ, and a vortex carries one fluxoid $\Phi = \hbar c/2e$, so $B_{max} \approx \Phi/\pi\xi^2$.

Solution 5.3. a) Since we know that the low-frequency linear vibrations are described by ordinary phonons, the specific heat will have the usual phonon term proportional to T^3. Let us calculate the specific heat $C_v(E_\alpha)$ due to one "two-level" center, i.e., a system of two energy levels separated by an energy E_α. We choose the zero of energy so that we may write the two energies in a symmetric fashion:

$$E_1 = \tfrac{1}{2}E_\alpha, \quad E_2 = -\tfrac{1}{2}E_\alpha. \qquad (14.22)$$

The internal energy is

$$U = \frac{\sum_\epsilon \epsilon e^{-\beta\epsilon}}{\sum e^{-\beta\epsilon}} = -\frac{E_\alpha}{2}\tanh\frac{E_\alpha\beta}{2}, \qquad (14.23)$$

where we summed over $\epsilon = E_1, E_2$, with $\beta \equiv 1/k_BT$. This is the usual result for a two-level system. The specific heat is

$$C_v(E_\alpha) = \frac{\partial U}{\partial T} = \frac{k_B}{4}E_\alpha^2\beta^2\mathrm{sech}^2\frac{E_\alpha\beta}{2}. \qquad (14.24)$$

We need to integrate the specific heat over the distribution of energies, $P(E_\alpha)$ (which we are told is a constant, P) to find the average value $\overline{C_v}$:

$$\overline{C_v} = \int_0^\infty dE\, Pk_B\left(\frac{E\beta}{2}\right)^2 \mathrm{sech}^2\left(\frac{E\beta}{2}\right). \qquad (14.25)$$

We make the substitution $x = E\beta/2$ to find

$$\overline{C_v} = 2IPk_B^2T, \qquad (14.26)$$

where $I = \int_0^\infty dx\,(x^2/\cosh^2 x) = \pi^2/12$ is an integral that can be found in books.

b) The specific heat, equation (14.26), has the same temperature dependence as the specific heat for a degenerate electron gas, such as in a metal: $C_v^e \propto T$ In order to estimate the relative magnitudes, we need the explicit form of C_v^e:

$$C_v^e = \left(\frac{\pi^2}{2}\right) N_e k_B \left(\frac{k_B T}{\epsilon_F}\right). \tag{14.27}$$

(Notice that the numerical factor is approximately 5, nearly an order of magnitude. The derivation of this formula may be found in many texts. See, for example, Ashcroft and Mermin, Chapter 2.) Here, N_e is the number of mobile electrons in the metal.

We can write the density of "two-level" centers per unit energy as $P = N_a g$, where N_a is the number of atoms in the sample and g is the density of centers per atom, which we are given is 1 state/(eV atom). Thus the ratio of specific heats is

$$\frac{\overline{C_v}}{C_v^e} = \frac{\epsilon_F N_a g}{3 N_e}. \tag{14.28}$$

The number of electrons in the metal is $N_e = m N_a$, where m is an integer usually between 1 and 3, corresponding to the number of mobile electrons per atom. Typically ϵ_F is between 5 and 10 eV for metals. With these numbers, we find that the ratio is close to unity. Thus the specific heats are of roughly equal magnitude.

c) This sample is probably graphite, which is composed of sheets of carbon atoms, so that it is nearly a two-dimensional system. To confirm this, we find the low temperature T-dependence of the specific heat for a two-dimensional phonon gas. Recall that, for a boson gas,

$$C_v \propto \frac{\partial}{\partial T} \int \frac{d\mathbf{k} \, \hbar\omega}{e^{\beta\hbar\omega} - 1}. \tag{14.29}$$

In two dimensions, $d\mathbf{k} = 2\pi k \, dk \propto \omega \, d\omega$. We rewrite the integral above in terms of the dimensionless parameter $x \equiv \hbar\omega\beta$, and find

$$C_v \propto \frac{\partial}{\partial T} \left[\frac{1}{\beta^3} \int \frac{x^2 dx}{e^x - 1}\right] \propto T^2. \tag{14.30}$$

This supports our guess that the crystalline carbon is graphite.

Solution 5.4. a) Since the $\hat{x}$, $\hat{y}$ and $\hat{z}$ crystal axes are interchangeable, the expansion of the free energy will be symmetric with respect to the components of the order parameter: P_x, P_y and P_z. In addition, there is no reason to favor, for example, a polarization along the $+\hat{x}$ direction as opposed to the $-\hat{x}$ direction. Thus, only even powers of the order parameter components will appear in the free energy f. If we assume there is no externally applied field, we can write (to fourth order)

$$f = A(P_x^2 + P_y^2 + P_z^2) + B(P_x^4 + P_y^4 + P_z^4) + C(P_x^2 P_y^2 + P_y^2 P_z^2 + P_z^2 P_x^2) ,$$
$$(14.31)$$

where A, B and C are (so far) arbitrary coefficients which depend on pressure and temperature.

The physical value of $\mathbf{P}$ will be at the global minimum of the free energy. In order for this minimum to occur at a finite value of $\mathbf{P}$, we require $B > 0$. By considering points along the line $P_x = P_y = P_z$ far from the origin, we can also set the condition that $C \geq -B$.

If the polarization is zero, the free energy will likewise be zero. In order for the free energy to have a minimum at a non-zero value of $\mathbf{P}$, we require $A < 0$ (below T_c), so that at small values of $\mathbf{P}$ the free energy is negative, and then at large values of $\mathbf{P}$ it turns and heads toward infinity.

b) In order to find the possible polarization vectors, we need to find the minima of f. The minima satisfy $\partial f / \partial P_i = 0$, which gives the equation

$$2AP_x + 4BP_x^3 + 2CP_x(P_y^2 + P_z^2) = 0 , \qquad (14.32)$$

as well as two more equations with $x \leftrightarrow y$ and $x \leftrightarrow z$, and we must solve these three equations.

First, if all the components of $\mathbf{P}$ are nonzero, then they must all be equal to satisfy simultaneously equation (14.32) and its permutations. Solving for $P_x = P_y = P_z$ and substituting into the free energy equation (14.31) gives

$$\text{(Case 1)} \quad P_x = P_y = P_z \Rightarrow f = -\frac{3}{4}\frac{A^2}{(B+C)} . \qquad (14.33)$$

Next, suppose one component is zero, say P_x. Then P_y and P_z must be equal and we find

$$\text{(Case 2)} \quad P_x = 0 \Rightarrow P_z = P_y \Rightarrow f = \frac{-A^2}{2B+C} \, . \qquad (14.34)$$

Finally, suppose two components are zero, say P_x and P_y:

$$\text{(Case 3)} \quad P_x = P_y = 0 \Rightarrow f = \frac{-A^2}{4B} \, . \qquad (14.35)$$

Clearly, in cases (2) and (3), there is nothing special about the choice of indices x, y and z, and any permutation of them also gives a solution with the corresponding free energy. In addition, we can change the sign of any of the components (e.g., $P_x \rightarrow -P_x$) and we still have a solution. In order for a polarization to appear, it must have a free energy lower than that of any of the other possible polarization states. With this requirement one can show that case (2) never appears. As the coefficient A passes through zero to become negative, the resulting spontaneous polarization will depend on the values of B and C. The possible polarizations are:

$$\text{(Case 1)} \quad -B < C < 2B \Rightarrow \quad \mathbf{P} = P(\hat{\mathbf{x}} + \hat{\mathbf{y}} + \hat{\mathbf{z}}), \quad (14.36)$$
$$\text{(Case 3)} \quad C > 2B \quad \Rightarrow \quad \mathbf{P} = P\hat{\mathbf{z}}, \qquad (14.37)$$

Regardless of which polarization develops, the transition is second order. Of course, the specific choice of the cube diagonal used in case (1) or the principal axis in case (3) is arbitrary (in the absence of an ordering field).

c) A second order phase transition is characterized by a continuous symmetry change, like the slow growth of a polarization from zero to a finite value. A first order transition has a finite change in the order parameter, for example, a direction change of the polarization. A possible phase diagram is shown in Figure 14.3. To draw the real phase diagram we would need to know the pressure and temperature dependence of the coefficients A, B and C. In the figure we have (arbitrarily) chosen $C < 2B$ for large P and T. When B and C, which change slowly with pressure and temperature, have evolved into the regime $C > 2B$,

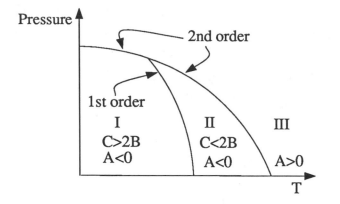

Figure 14.3: A plausible phase diagram. I and II designate different polarizations, $\mathbf{P}_I = P\hat{\mathbf{z}}$ and $\mathbf{P}_{II} = P(\hat{\mathbf{x}}+\hat{\mathbf{y}}+\hat{\mathbf{z}})$. There is no polarization in region III.

the polarization will suddenly switch from the form in case (1) to that in case (3), provided $A < 0$, and therefore there is a first-order phase transition along the path $C = 2B$ in P-T space.

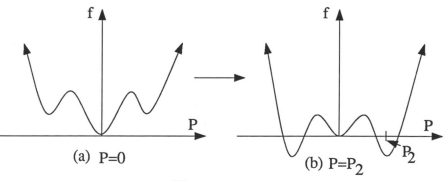

Figure 14.4.

d) We must first deduce the contributions of the strain to the free energy. To make things simple, consider only distortions in the z-direction, as shown in Figure 5.5. The strain s is the fractional change in length of the crystal; there must be an energy associated with this stretching, which should look like that of a spring, or

$$f_{strain} = \frac{1}{2}ks^2 \ . \tag{14.38}$$

The problem suggests that there is a coupling term between the polarization and the strain. The simplest possible coupling will be linear in the strain and have two powers of the polarization, since the coupling should not depend on the direction of the polarization. Thus the coupling f_c will have the form $f_c = DsP_z^2$, so that the terms due to the strain add up to

$$f_s = \frac{1}{2}ks^2 + DsP_z^2 \; . \tag{14.39}$$

To find the minimum of f_s, we differentiate with respect to s, yielding

$$f_s = -\frac{D^2P_z^4}{2k} \; . \tag{14.40}$$

This term adds to those in the previous free energy expansion (14.31), and if k is small enough (i.e., the crystal is "soft" enough), f_s will overwhelm the coefficients in front of the other P^4 terms in that expansion, and make the effective coefficient of P^4 negative. In order to keep the minimum of the free energy at a finite value of P, we have to add P^6 terms (with positive coefficients). Note that a nonzero polarization can now arise even if the coefficient A is greater than zero, in which case a negative P^4 term causes the symmetry to change abruptly; the polarization jumps from zero (Figure 14.4a) to a finite value (Figure 14.4b). Thus the phase transition is first order. (Also see Huang, Chapter 17, for more details.)

Solution 5.5. a) Of course, strips of copper and tin alone will not go very far towards cooling anything. Add a voltage source as shown in Figure 14.5, however, and we can make use of the Peltier effect to transport heat from point A to point B, or vice versa. The Peltier effect is the name given to the fact that an electric current j_e in a conductor gives rise to a thermal current j_{th} which is proportional to j_e, with a constant of proportionality Π, the Peltier coefficient, which is characteristic of the conductor: $j_{th} = \Pi j_e$ (see Ashcroft and Mermin,

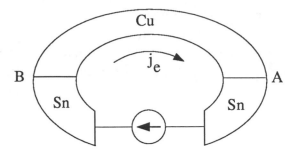

Figure 14.5.

Chapter 13). In the circuit shown, the electric current j_e is constant around the circuit, so the thermal current across the device is

$$j_{th}(B \to A) = (j_{th}^{Cu} - j_{th}^{Sn}) = (\Pi_{Cu} - \Pi_{Sn})j_e \, . \qquad (14.41)$$

b) EuO is probably a ferromagnet. We recognize the T^3 term in the heat capacity as the standard Debye result for the contribution from phonons in a three-dimensional lattice at temperatures low compared to the lattice Debye temperature. The $T^{3/2}$ term is the contribution from magnons (i.e., spin waves), the elementary excitations of a ferromagnet. To derive the $T^{3/2}$ temperature dependence, recall that magnons are bosons, with an energy-momentum dispersion relation $E \propto k^2$ (see Ziman, Chapter 10), as opposed to $E \propto k$ for phonons. Then it is straightforward to show that $C \propto T^{3/2}$ by proceeding as one would to find, say, the phonon contribution to the specific heat.

c) V_3Ga is evidently a type II superconductor: an applied magnetic field can penetrate it, with the lines of flux confined to a lattice of non-superconducting fluxoids, up to a maximum value of the applied field (which is shown in Figure 5.6 to be 3 kgauss) beyond which the metal ceases to be a superconductor. Each fluxoid carries one quantum of flux, which has the value $\Phi_0 = hc/2e = 2.1 \times 10^{-7}$gauss cm^2. According to Figure 5.6, a field $B \approx 2.3 \times 10^3$ gauss penetrates the superconductor. Since this field is near the saturation value, we can guess that the fluxoids will be arranged in a hexagonal close-packed pattern. The surface area per fluxoid is $A = \sqrt{3}d^2/2$, where d is the distance between fluxoids. Then $B = \Phi_0/A$ gives $d \approx 10^{-5}$ cm.

d) Let θ be the relative angle between the beam path and the symmetry axis of a single crystal. Then the Bragg scattering condition, $n\lambda = 2b\sin\theta$, where n is an integer, will determine which neutrons are scattered out of the beam path into the walls of the pipe. In a powder the angle θ will vary randomly from 0 to $\pi/2$, so all neutrons with $\lambda < 2b$ will be scattered. Any neutron with $\lambda > 2b$ will be transmitted through the pipe.

e) This is the standard geometry of a Hall effect experiment. Let the average drift velocity of an electron in the direction opposite the current be $\mathbf{v}$. Then the current density is $\mathbf{J} = -ne\mathbf{v}$ in the direction shown in Figure 5.7. (Note that the total current is $I = Ja$, where a is the cross-sectional area of the slab.) The electrons will feel a Lorentz force given by $F = -e\mathbf{v} \times \mathbf{B}$ (in SI units), and since no current flows out of the sides of the strip, there must be an electric field induced in the sample that exactly counterbalances this force. In terms of the measured voltage, the magnitude of the induced field is $V/(10 \text{ mm})$. Setting the two forces equal yields

$$\frac{V}{10 \text{ mm}} = |\mathbf{v}|B = \frac{IB}{ane}. \tag{14.42}$$

To use SI units we must convert B to tesla ($1 \text{ T} = 10^4$ gauss), and use units of meters for all lengths. This gives us $n = 7.7 \times 10^{22} \text{ cm}^{-3}$.

f) The mechanism by which electrons incident on a thin film of magnesium lose energy is the production of surface plasmons. The existence of surface waves on a metal, with frequency $w_s = w_p/\sqrt{2}$, can be shown classically. Here, the plasma frequency w_p is given by $w_p^2 = 4\pi ne^2/m$ (see Ashcroft and Mermin, Chapter 1). After quantization, these waves become boson excitations with energy $\hbar w_s$. The incident electron can excite an integral number of these plasmons, so the energy-loss spectrum will have peaks at $E = \hbar w_s, 2\hbar w_s, \ldots$ (Figure 14.6).

Solution 5.6. a) The density of states is given by $\rho(E) \propto q^2 |dq/dE|$, where the right-hand side is written as a function of energy E. To

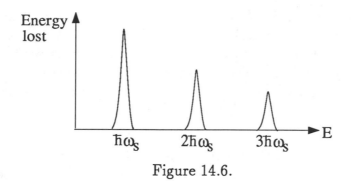

Figure 14.6.

find $|dq/dE|$, we consider separately the three regions given by $E < \Delta$, $\Delta < E < \epsilon$ and $E > \epsilon$. For $E < \Delta$, $E \approx qc$, so that $\rho(E)$ increases as E^2. When E reaches Δ from below, the density of states jumps discontinuously due to the new contribution from the roton states. In fact, $|dq/dE|$ (and therefore ρ) is infinite at $E = \Delta$. As E approaches Δ from above, $|dq/dE| \approx 1/(2\sqrt{\alpha(E-\Delta)}+$constant, so that the density of states rises smoothly to infinity. Similar behavior occurs near $E = \epsilon$. This density of states is sketched in Figure 14.7.

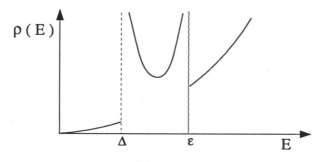

Figure 14.7.

b) For wavevectors near q_0, the dispersion relation is

$$E \approx \Delta + \alpha k^2, \tag{14.43}$$

where $k = q - q_0$. If we define $m \equiv \hbar^2/2\alpha$, then by analogy with the usual nonrelativistic energy-momentum relationship $E = \hbar^2 k^2/2m$, equation (14.43) describes a free quasiparticle of mass m in a constant potential Δ: a roton. Let us write a single roton wavefunction as

$e^{i\mathbf{r}\cdot\mathbf{q}_0}\phi(\mathbf{r})$. By inspection of (14.43), the action of the hamiltonian H on this wavefunction is

$$H e^{i\mathbf{r}\cdot\mathbf{q}_0}\phi(\mathbf{r}) = e^{i\mathbf{r}\cdot\mathbf{q}_0}\left(\frac{\hbar^2}{2m}\nabla_r^2 + \Delta\right)\phi(\mathbf{r}) = E e^{i\mathbf{r}\cdot\mathbf{q}_0}\phi(\mathbf{r}),\qquad (14.44)$$

where E is the energy of the roton and ∇_r^2 is the Laplacian with respect to $\mathbf{r}$. Therefore the hamiltonian must be

$$H \equiv e^{i\mathbf{r}\cdot\mathbf{q}_0}\left(\frac{\hbar^2}{2m}\nabla_r^2 + \Delta\right)e^{-i\mathbf{r}\cdot\mathbf{q}_0}.\qquad (14.45)$$

In using this hamiltonian, we are assuming that the wave packet $\phi(\mathbf{r})$ only has Fourier components at wavevectors near zero, i.e., $k \approx 0$, where the dispersion relation (14.43) is valid.

Now let us introduce the contact interaction $V(\mathbf{r}_1-\mathbf{r}_2)$, and consider two rotons each with wavevectors near $\mathbf{q}_0$ and with position operators $\mathbf{r}_1$ and $\mathbf{r}_2$. In the center-of-mass coordinates given by $\mathbf{R} = (\mathbf{r}_1 + \mathbf{r}_2)/2$ and $\mathbf{r} = \mathbf{r}_1 - \mathbf{r}_2$, the Schrödinger equation for the wavefunction $e^{2i\mathbf{q}_0\cdot\mathbf{R}}\Phi$ of the two rotons is

$$e^{i2\mathbf{q}_0\cdot\mathbf{R}}\left(-\frac{\hbar^2}{4m}\nabla_R^2 - \frac{\hbar^2}{2\mu}\nabla_r^2 + V(\mathbf{r})\right)\Phi = (E - 2\Delta)\Phi e^{i2\mathbf{q}_0\cdot\mathbf{R}},\quad (14.46)$$

where $\mu = m/2$, and the total energy of the two-roton system is E. Clearly Schrödinger's equation is separable and we can write the solution as $\Phi(\mathbf{r},\mathbf{R}) = \exp(i\mathbf{K}\cdot\mathbf{R})\psi(\mathbf{r})$. We will only consider states with $\mathbf{K} = 0$. In this case, equation (14.46) reduces to

$$\left(-\frac{\hbar^2}{2\mu}\nabla_r^2 + V(\mathbf{r})\right)\psi = (E - 2\Delta)\psi.\qquad (14.47)$$

A bound state will occur if there exists a solution with $E < 2\Delta$. If we put the superfluid in a box of volume Ω, we can expand ψ in a basis of normalized plane-wave states as $\psi = (1/\sqrt{\Omega})\sum_k A_k e^{i\mathbf{k}\cdot\mathbf{r}}$. Taking the Fourier transform of the Schrödinger equation (14.47) yields

$$\left(E_B - \frac{\hbar^2 k^2}{2\mu}\right)A_k = \sum_{k'} V_{kk'}A_{k'},\qquad (14.48)$$

where $E_B \equiv E - 2\Delta$, and the Fourier components of the potential are

$$V_{\mathbf{k}\mathbf{k}'} = -\frac{g}{\Omega} \int e^{i\mathbf{r}\cdot(\mathbf{k}'-\mathbf{k})} \delta^3(\mathbf{r}) d^3r = -\frac{g}{\Omega} . \tag{14.49}$$

(In the above, each sum over $\mathbf{k}$ or $\mathbf{k}'$ has an implicit cutoff so that the dispersion relation (14.43) remains valid.) Substituting $V_{\mathbf{k}\mathbf{k}'}$ into the transformed Schrödinger equation (14.48) gives

$$A_{\mathbf{k}} = \left(-\frac{g}{\Omega}\right) \frac{\sum_{\mathbf{k}'} A_{\mathbf{k}'}}{E_B - \hbar^2 k^2/2\mu} . \tag{14.50}$$

Summing both sides over $\mathbf{k}$ up to the cutoff, one finds

$$-\frac{\Omega}{g} = \sum_{k} \frac{1}{E_B - \hbar^2 k^2/2\mu} , \tag{14.51}$$

after canceling common factors of $\sum_{\mathbf{k}} A_{\mathbf{k}}$. In order to satisfy this equation for small g and finite negative E_B (and consequently in order for a bound state to arise), there must be terms in the sum with $\mathbf{k} \approx 0$. This requires that the density of states for the rotons be nonzero at $\mathbf{k} = 0$.

For a large volume Ω, we may approximate the sum on $\mathbf{k}$ as an integral:

$$\sum_{k} \rightarrow \frac{\Omega}{(2\pi)^3} \int d^3q , \tag{14.52}$$

where $\mathbf{q} = \mathbf{q}_0 + \mathbf{k}$. The eigenvalue equation (14.51) becomes

$$\frac{1}{g} = \frac{1}{2\pi^2} \int_{-\delta}^{\delta} \frac{(q_0 + k)^2 dk}{-E_B + 2\alpha k^2} . \tag{14.53}$$

We have now explicitly instated the cutoff δ on k. If there exists an $E_B < 0$ for which (14.53) holds, then a bound state exists. We approximate the numerator in the integral (14.53) by q_0^2, and make the substitution $x = k\sqrt{-2\alpha/E_B}$, to find

$$-\frac{1}{g} \approx \frac{q_0^2}{2\pi^2} \frac{2}{\sqrt{-2\alpha E_B}} \int_0^{\sqrt{-2\alpha\delta^2/E_B}} \frac{dx}{1 + x^2} . \tag{14.54}$$

This integrates to give

$$\frac{1}{g} = \frac{q_0^2}{\pi^2} \frac{1}{\sqrt{-2\alpha E_B}} \left[\arctan\left(\sqrt{\frac{-2\alpha}{E_B}} \delta \right) \right] . \tag{14.55}$$

If we assume that the binding energy is small, since we are interested in the limit of small g, then the argument of the arctangent above is very large for any finite δ. Since $\arctan \infty = \pi/2$, we arrive at the approximate expression,

$$\frac{1}{g} \approx \frac{q_0^2}{2\pi} \frac{1}{\sqrt{-2\alpha E_B}} , \qquad (14.56)$$

which is independent of the cutoff. Rearranging to solve for E_B, gives

$$E_B \approx -\frac{g^2 q_0^4}{8\pi^2 \alpha} < 0 . \qquad (14.57)$$

Therefore there is no critical value of g: a bound state always exists.

The reader may recall that for a particle in a three-dimensional vacuum in the presence of a delta-function potential, there is a critical value of g for the appearance of a bound state. This is not inconsistent with the result above because the case of a single particle (in notation corresponding to that used in this problem) has $q_0 = 0$, which means that the density of states is zero at $k = 0$, and (14.57) yields $E_B = 0$. In one dimension, on the other hand, there is no critical value of g for a single particle because the density of states at $k = 0$ does not vanish.

The binding of rotons is closely related to the phenomenon of Cooper pairs, which are the bound pairs of electrons in a degenerate Fermi electron gas that are responsible for superconductivity (although rotons are not responsible for superfluidity).

Solution 5.7. a) The elastic forces are trying to keep the molecules lined up in the x-direction, so for any interesting physics to occur, it must be that the magnetic field is trying to line them up in the y-direction (we will see this later). Take l to be the characteristic length in the z-direction over which θ is changing rapidly (as shown in Figure 14.8). For the system to be at a minimum of the total free energy, the derivatives of the magnetic and elastic energies with respect to l must be equal and opposite.

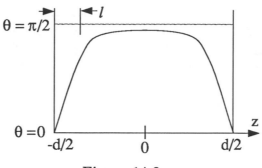

Figure 14.8.

The magnetic free energy density is given by

$$f_{mag} = -\frac{1}{2}\mathbf{H}\bar{\chi}\mathbf{H} \;\; = \;\; -\frac{1}{2}(\chi_0 H^2 + \chi_a H^2 \sin^2\theta) \;, \qquad (14.58)$$

$$= \;\; -\frac{1}{2}(\chi_0 + \chi_a)H^2 + \frac{1}{2}\chi_a H^2 \cos^2\theta \;,$$

in analogy with $df = -MdH = -\chi HdH = d(-\chi H^2/2)$. Note that θ is the angle the molecules make with the $\hat{x}$-axis, not the angle they make with the field. Since $\cos\theta \approx 0$ except in the region where θ is varying rapidly, the total magnetic free energy is approximately

$$E \sim \chi_a H^2 Al + C \;, \qquad (14.59)$$

where A is the face-on area of the glass slides, and C is independent of l.

By dimensional analysis, the total elastic free energy (5.2) is $E \sim kAl/l^2$, where we have replaced the derivative with respect to length by $(1/l)$. Equating the magnitudes of the derivatives (with respect to l) of these two expressions gives the characteristic length scale,

$$l = \left(\frac{k}{\chi_a H^2}\right)^{\frac{1}{2}} \;. \qquad (14.60)$$

b) The unit vector $\hat{n}$ will distort from its zero-field direction when the energy cost of doing so is less than the energy gained by aligning the molecules with the field (provided the boundary conditions at the glass slides can be met). Combining the given expression for the elastic energy (5.2) with the magnetic field energy density (14.58) gives the total free energy density f:

$$f = \frac{1}{2}k \sum_{i,j} \left(\frac{\partial n_j}{\partial x_i}\right)^2 - \frac{1}{2}(\chi_0 H^2 + \chi_a H^2 \sin^2 \theta) . \qquad (14.61)$$

Using the given form for $\hat{n}(z)$, equation (5.3), we find

$$\sum_{i,j} \left(\frac{\partial n_j}{\partial x_i}\right)^2 = \left(\frac{\partial \theta}{\partial z}\right)^2 . \qquad (14.62)$$

Substituting this quantity into (14.61) gives an expression for the free energy density, in terms of θ and $\dot{\theta} \equiv \partial \theta / \partial z$:

$$f = \frac{k}{2}\dot{\theta}^2 - \frac{\chi_0}{2}H^2 - \frac{\chi_a}{2}H^2 \sin^2 \theta . \qquad (14.63)$$

Using Lagrange's equation we find

$$\ddot{\theta} + \frac{\chi_a H^2}{k} \sin \theta \cos \theta = 0 . \qquad (14.64)$$

Since we expect a small distortion, we use $\sin \theta \approx \theta$, and $\cos \theta \approx 1$, which leads to the solution

$$\theta = b \cos \left(\sqrt{\frac{\chi_a H^2}{k}} z\right) , \qquad (14.65)$$

with b some constant.

This expression for θ is only correct if the boundary conditions that $\theta = 0$ at each slide (positioned at $z = \pm d/2$) can be met. This yields the result that the first (small) deviation from $\theta = 0$, at the center of the liquid, will occur when

$$H_c = \frac{\pi}{d} \left(\frac{k}{\chi_a}\right)^{\frac{1}{2}} . \qquad (14.66)$$

The amplitude b of the distortion is not determined by the equations in the small-θ approximation. In order to find b we would have to solve the non-linear differential equation (14.64).

c) Integrating equation (14.64) with respect to z yields a conserved quantity, which we write as

$$\dot{\theta}^2 + \frac{\chi_a H^2}{k} \sin^2 \theta = \frac{\chi_a H^2}{k} \sin^2 \theta_{max} , \qquad (14.67)$$

since $\theta = \theta_{max}$ when $\dot{\theta} = 0$ (which occurs at $z = 0$, by inspection). We can integrate over half the distance between the slides to find

$$\int_0^{\theta_{max}} d\theta \frac{1}{\sqrt{\sin^2 \theta_{max} - \sin^2 \theta}} = \int_{-d/2}^0 \frac{dz}{l} , \qquad (14.68)$$

where we have defined $l^2 \equiv k/(\chi_a H^2)$. The integral on the left hand side is dominated by the region $\theta \approx \theta_{max}$, where the denominator vanishes. If we write $\theta = \pi/2 - \delta$, so that $\sin^2 \theta \approx 1 - \delta^2$, equation (14.68) becomes

$$\int_{\delta_0}^{\pi/2} \frac{d\delta}{\sqrt{\delta^2 - \delta_0^2}} \approx \frac{d}{2l} , \qquad (14.69)$$

where $\sin^2 \theta_{max} = 1 - \delta_0^2$. Substituting $\delta = \delta_0 \cosh u$ leads to

$$\delta_0^{-1} \approx \frac{2}{\pi} \cosh \left(\frac{d}{2l} \right) . \qquad (14.70)$$

If H is very large, $d/2l$ will also be very large. Since $\cosh(x) \approx (1/2)e^x$ in the limit $x \gg 1$, we can see that (14.70) gives the maximum distortion as

$$\delta \approx \pi e^{-d/2l} \propto e^{-(\pi/2)(H/H_c)} , \qquad (14.71)$$

where H_c is the critical field defined in equation (14.66).

Solution 5.8. a) We need to consider three contributions to the energy of the electron in the lattice. The first is the potential $V(\mathbf{x}) = \epsilon_0 \nabla \cdot \mathbf{U}$, which we are taking to be a constant within the volume occupied by the electron (and zero outside). The second is the elastic energy of the lattice distortion: if the bulk modulus is B then, for a distortion $\nabla \cdot \mathbf{U}$, the energy per unit volume is $\frac{1}{2}B(\nabla \cdot \mathbf{U})^2$. Thus the total energy of a distortion of linear size L is

$$E_D = \int \frac{1}{2}B(\nabla \cdot \mathbf{U})^2 \, dV = \frac{1}{2}B(\nabla \cdot \mathbf{U})^2 L^d. \tag{14.72}$$

The last contribution is the kinetic energy of the electron. For simplicity, we will find the approximate kinetic energy by considering the electron to be confined in an infinitely deep square well potential of size L^d. Then the lowest energy eigenstate has components of momentum given by $k_i = \pi/L$, and the kinetic energy in d dimensions is

$$T = \frac{\hbar^2 k^2}{2m} = \frac{d\hbar^2 \pi^2}{2mL^2}. \tag{14.73}$$

The total energy is then

$$E(L) = \frac{d\hbar^2 \pi^2}{2mL^2} + \frac{1}{2}B(\nabla \cdot \mathbf{U})^2 L^d + \epsilon_0 \nabla \cdot \mathbf{U}. \tag{14.74}$$

We minimize this with respect to $V_0 = \epsilon_0 \nabla \cdot \mathbf{U}$ for fixed L:

$$\frac{\partial E}{\partial V_0} = \frac{BL^d V_0}{\epsilon_0^2} + 1 = 0. \tag{14.75}$$

So $V_0 = -\epsilon_0^2/BL^d$, or

$$\nabla \cdot \mathbf{U} = -\epsilon_0/BL^d. \tag{14.76}$$

It is easy to check that this is a minimum.

b) We can use the optimum value for the potential, equation (14.76), in the expression (14.74) for the energy to write E as a function of L only:

$$E(L) = \frac{d\hbar^2 \pi^2}{2mL^2} + \frac{\epsilon_0^2}{2BL^d} - \frac{\epsilon_0^2}{BL^d} = \frac{d\hbar^2 \pi^2}{2mL^2} - \frac{\epsilon_0^2}{2BL^d}. \tag{14.77}$$

The competition between these two terms decides the fate of the polaron. The second term, the sum of the elastic and potential energies, favors small deformations. However, the kinetic energy term reflects the antipathy of a quantum-mechanical particle towards being confined in a small region: the outcome, in the limit as L becomes small, will depend on the dimension d. Clearly, for $d < 2$, $E(L) \to +\infty$ as $L \to 0$, whereas for $d > 2$ we get $E(L) \to -\infty$ in the same limit. As L increases, there will in both cases be a value L_0 for which $E(L_0) = 0$, given by

$$L_0^{d-2} = \frac{m\epsilon_0^2}{dB\hbar^2\pi^2}, \tag{14.78}$$

and eventually $E(L)$ will tend to zero as $L \to \infty$. For $d = 1$ we see

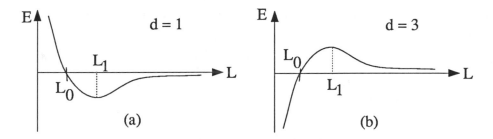

Figure 14.9.

that there is a well-defined minimum of $E(L)$, which will correspond to a stable polaron of size L_1, as shown in Figure 14.9a. For $d = 3$, the energy can become arbitrarily negative as the polaron shrinks. In this case, the energy has a maximum at $L = L_1$, as shown in Figure 14.9b.

c) We now take into account the discrete nature of the lattice, which imposes a natural cut-off on the size of a polaron. We assume that the elastic energy has the following form:

$$E_D = \begin{cases} \frac{1}{2}B(\boldsymbol{\nabla}\cdot\mathbf{U})^2 L^d & L > a, \\ \infty & L < a. \end{cases} \tag{14.79}$$

The polaron will now be the true ground state if there exists a value of L such that $L > a$ and $E(L) < 0$. In the case of $d = 3$, we require $L_0 > a$, or

$$\epsilon_0^2 > \frac{3aB\hbar^2\pi^2}{m}. \tag{14.80}$$

(If this is not satisfied, then it is still possible to get metastable polaronic states, with $L_0 \approx a$, provided $L_1 > a$. However these states are unstable with respect to quantum-mechanical tunneling through the potential barrier.)

d) In $d = 1$ there is always a polaron, since for any $L > L_0$ the energy is negative; the only variable is its size. If $L_1 > a$ then the polaron will sit at the minimum of the potential and have size L_1. If $L_1 < a$ the polaron will be as small as the lattice permits, namely of order a.

———

Solution 5.9. a) i) At zero temperature the electrons fill every energy state up to the Fermi energy $\epsilon_f = \hbar^2 k_f^2 / 2m$. Since there is one state in each interval of length $(2\pi/L)$ in k-space, we can write the total kinetic energy U as

$$U = \frac{L}{2\pi} \int_{-k_f}^{k_f} dk \frac{\hbar^2 k^2}{2m} = \frac{\hbar^2 k_f^3 L}{6\pi m} = \frac{\pi^2 \hbar^2 n^3}{6m} L, \qquad (14.81)$$

where $n \equiv N/L$ and we have used $k_f = N\pi/L$ in the last equality.

ii) For a paramagnetic system, we gain a factor of 2 in front of the sum for U from the spin degeneracy; however, k_f is now half of its previous value for the same reason, so that the new kinetic energy is

$$U = \frac{\pi^2 \hbar^2 n^3}{24m} L. \qquad (14.82)$$

b) Substituting $V(q) = G$ into the expression for V_{HF} (5.6) and summing over k' and s' gives

$$V_{HF} = \frac{G}{2} \left\{ \sum_{ks} (N - N_s) c_{ks}^\dagger c_{ks} \right\}, \qquad (14.83)$$

where $N_s = \sum_k n_{ks}$ is the total number of electrons with spin s. The single-particle energies are simply

$$\epsilon_{ks}' = \frac{\hbar^2 k^2}{2m} + \frac{1}{2} G(N - N_s). \qquad (14.84)$$

c) In the ferromagnetic state, $N = N_s$ so that the single-particle energies are the same as in part (a) and therefore the ground state energy is also the same:

$$E^{HF}(\text{ferro}) = \frac{\pi^2 \hbar^2 n^3}{6m} L. \tag{14.85}$$

In the paramagnetic case, $N_s = N/2$, so that

$$E^{HF}(\text{para}) = \frac{\pi^2 \hbar^2 n^3}{24m} L + \frac{1}{4} G n^2 L^2. \tag{14.86}$$

The difference between these two ground state energies is then

$$\frac{\Delta E}{N} = \frac{\pi^2 \hbar^2 n^2}{8m} - \frac{1}{4} G L n, \tag{14.87}$$

which is plotted in Figure 14.10. The critical density is the density (for

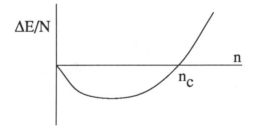

Figure 14.10.

a fixed length) at which it is energetically more favorable for the gas to be in a paramagnetic state than a ferromagnetic one. In other words, the transition is at $\Delta E/N = 0$, which occurs at the critical density, $n_c = 2GLm/\pi^2 \hbar^2$.

Solution 5.10. (This problem sketches the physics behind the integer quantum Hall effect.)

a) The time-independent Schrödinger equation for an electron of charge $-e$ and effective mass m^* in a magnetic field is

$$\frac{1}{2m^*}\left(-i\hbar\nabla + \frac{e}{c}\mathbf{A}\right)^2\psi + \frac{g_s\mu_B}{2}\boldsymbol{\sigma}\cdot\mathbf{B}\psi = E\psi. \qquad (14.88)$$

If we consider the term linear in $\mathbf{A}$, it is simple to see that the assumption $g_s \ll m/m^*$ is just what we need in order to neglect the interaction of the electron spin with the magnetic field, e.g.,

$$\left|\frac{g_s e\hbar}{4mc}\boldsymbol{\sigma}\cdot\mathbf{B}\right| \ll \left|\frac{e\hbar}{2m^*c}\mathbf{A}\cdot\nabla\right|. \qquad (14.89)$$

If we use the gauge suggested, $\mathbf{A} = (0, Bx)$, equation (14.88) becomes

$$\frac{1}{2m^*}\left(p_x^2 + \left[p_y + \frac{eBx}{c}\right]^2\right)\psi = E\psi, \qquad (14.90)$$

with $p_x = -i\hbar(\partial/\partial x)$ and similarly for p_y. First note that since y does not appear in the hamiltonian, p_y commutes with H and is therefore conserved. This means we can write the wavefunction as

$$\psi(x,y) = e^{ik_y y}\phi(x), \qquad (14.91)$$

with $\hbar k_y$ the wavevector in the y-direction. In terms of this wavefunction the Schrödinger equation is

$$\frac{1}{2m^*}\left(p_x^2 + \left[\hbar k_y + \frac{eBx}{c}\right]^2\right)\phi = E\phi, \qquad (14.92)$$

which is just the equation of motion of a one-dimensional simple harmonic oscillator with mass m^* and angular frequency $\omega = eB/m^*c$, oscillating about the point

$$x_0 = -\frac{\hbar c k_y}{eB}. \qquad (14.93)$$

The energies are $E_n = \hbar\omega(n + 1/2)$. We can write down an explicit wavefunction for the lowest energy states, which have $n = 0$. Recall that the ground-state wavefunction for a simple harmonic oscillator takes the form of a Gaussian wave packet, $\phi(x) = A \exp[-a(x - x_0)^2]$. We substitute this $\phi(x)$ into equation (14.92) to find $a = eB/2\hbar c$. Thus the complete wavefunction is given by

$$\phi(x, y) = A \exp\left[ik_y y - \frac{eB}{2\hbar c}(x - x_0)^2\right], \qquad (14.94)$$

where A is a normalization constant. For each k_y, we have a wavefunction centered on $x = x_0(k_y)$, traveling in the y-direction.

For a *large* system, we can compute the degeneracy of each level via the semiclassical rule that each state occupies an area 2π in phase space, $\Delta k_y \Delta y = 2\pi$. Since the position (of the center of the wavefunction) in the x-direction is determined in terms of the y-momentum by equation (14.93), we may solve for the area of one state in configuration space:

$$\Delta x \Delta y = \frac{2\pi \hbar c}{eB}. \qquad (14.95)$$

We include a factor of 2 to account for the spin degeneracy of the electron and find the number of states with energy E_n in an area $\mathcal{A}$ is

$$N = \frac{2\mathcal{A}}{\Delta x \Delta y} = \frac{\mathcal{A}eB}{\pi \hbar c}, \qquad (14.96)$$

which is the degeneracy of each level.

b) If we apply an external voltage V, we will have a charge $Q = CV$ induced in the surface layer of the silicon, and consequently the number of electrons in the two-dimensional gas will be $N_e = Q/e$.

In part (a) we found energy levels (called *Landau* levels) at energies $E_n = \hbar\omega(n+1/2)$, which can each hold N electrons, where N is defined above. Therefore the number of filled levels is

$$p = \frac{N_e}{N} = \frac{Q\pi \hbar c}{\mathcal{A}e^2 B}. \qquad (14.97)$$

In general, p won't be an integer, and the highest level will therefore be only partially filled. We will use n^* to denote the highest filled or partially filled level. Then, the Fermi energy is

$$\epsilon_F = E_{n^*} = \hbar\omega(n^* + \frac{1}{2}). \qquad (14.98)$$

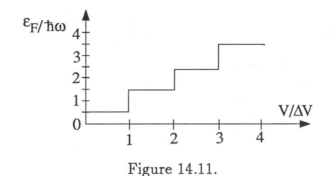

Figure 14.11.

Since we have a system of discrete energy levels, the Fermi energy will be a discontinuous function of the number of electrons in the surface layer (and consequently of the applied voltage): as soon as we have one electron in a new Landau level, the Fermi energy jumps up to its new value, and it stays there until that level is filled. The change in voltage required to fill one Landau level is

$$\Delta V = \frac{eN}{C} = \frac{\mathcal{A}e^2 B}{C\pi\hbar c}.$$ (14.99)

The Fermi energy is shown in Figure 14.11.

A small amount of disorder has two effects. The first is that the energy of each Landau level becomes smeared out over a small range so that the density of states is no longer comprised of delta functions at each $E_n = \hbar\omega(n+1/2)$, but is rather as shown in Figure 14.12. We will assume that the energy levels remain distinct, as pictured, and that the total number of states in each level is unchanged. The second effect of the disorder is the presence of completely new states in which electrons are bound to specific locations in the crystal. These locations mark places in the crystal where the regular order is disrupted, for example, by impurity atoms or vacant sites. The wavefunction of an electron in one of these states falls off rapidly with distance, and for this reason they are called *localized states*. (In contrast, electrons in the Landau levels can be called *delocalized*.) There are two important points about these states. The first is that because they are localized, they do not contribute to the current. The second point is that they have energies that lie in the gaps between the Landau energies. As a result, the Fermi

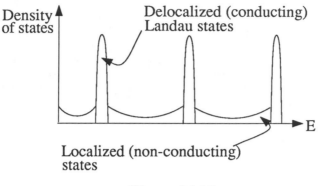

Figure 14.12.

energy now varies continuously as the number of electrons in the surface layer increases. When the Fermi energy falls within one of the peaks on the density of states, its rate of increase with applied voltage will be small: when it reaches one of the troughs between the old Landau levels, a small change in voltage will quickly fill the small number of available localized states and there will be a large rate of increase in the Fermi energy with applied voltage, as shown in Figure 14.13.

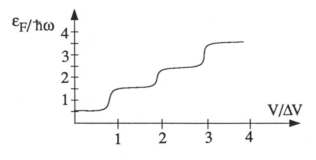

Figure 14.13.

c) An important feature of the Hall conductance is that *all* Landau levels contribute to the current. (This contrasts with the situation usually encountered in condensed matter physics, where only partially filled bands contribute.)

We start by noting that in crossed electric and magnetic fields, all electrons will drift with mean velocity $v = cE_H/B$ in the direction perpendicular to **E** and **B**. If we treat the metal as a semiclassical gas of electrons, the resulting current will be

$$j = \rho_d \, e v \,, \tag{14.100}$$

where ρ_d is the number of delocalized electrons per unit area.

The density of delocalized electrons is a function of the Fermi energy. We write $\rho_d = pN/\mathcal{A}$, where N is the degeneracy of each Landau level, which we found in part (a), and p is the (non-integral) number of filled levels, given in equation (14.97), which is a function of the Fermi energy. Then equation (14.100) gives

$$j = \frac{pN}{\mathcal{A}} e \frac{cE_H}{B} = \frac{pe^2 E_H}{\pi \hbar} \,. \tag{14.101}$$

Thus the Hall conductance is

$$\sigma_H = \frac{j}{E_H} = \frac{pe^2}{\pi \hbar} \,. \tag{14.102}$$

If the Fermi surface coincides with one of the Landau levels, the conductance will increase rapidly with increasing Fermi energy. However, if the Fermi surface lies between the Landau levels, in a region of energy where we have only localized states, increasing the Fermi energy does not increase the number of conducting electrons and the conductance shows a plateau at integer multiples of $e^2/(\pi \hbar)$ (which is known as the integer quantum Hall effect). This is shown in Figure 14.14.

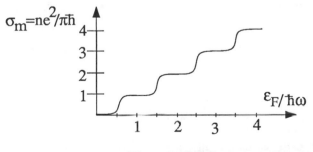

Figure 14.14.

Chapter 15

Relativity & Astrophysics—Solutions

Solution 6.1. a) For a massive particle, $E = \gamma mc^2$ and $E^2 = m^2 c^4 + p^2 c^2$. For the muon we are considering, $p \gg mc$, so that $\gamma = E/(mc^2) \approx p/(mc)$. A simple time dilation then gives the observed halflife as

$$t_{1/2} = \gamma \tau_{1/2} \approx \frac{p}{mc} \tau_{1/2} , \tag{15.1}$$

where $\tau_{1/2}$ is the halflife of the muon in its rest frame.

b) The average velocity of the nuclei will be given by the equipartition theorem, $mv^2/2 \approx 3kT/2$. The time dilation associated with this velocity will shift the observed resonance frequency ν_{lab} of the nuclei from its rest-frame value ν_{rest}, such that

$$\nu_{lab} = \frac{1}{\gamma} \nu_{rest} . \tag{15.2}$$

Inserting the average velocity leads to a shift given by

$$\nu_{lab} = \nu_{rest} \left(1 - \frac{3kT}{2mc^2} \right) . \tag{15.3}$$

c) The easiest way to calculate this frequency shift is to consider a thought experiment proposed by Einstein. Imagine that you have a magic machine that can change the energy of a photon into a mass, and the energy of a mass (plus its kinetic energy) into a photon. Then, put one of these machines at points A and B in Figure 15.1. If we begin by dropping a mass m from A to B, it gains energy mgH in falling to B. The machine at B then converts all that energy into a photon (with energy $h\nu_B$). This photon travels up to A, where it then has energy $h\nu_A$. If $h\nu_A$ were greater than mc^2, we could build a machine that would create energy from nothing. If $h\nu_A$ were less than mc^2, we would only have to operate the machine in reverse to create energy from nothing. Thus, $mc^2 = h\nu_A$, and

$$\frac{mc^2}{mc^2 + mgH} = \frac{h\nu_A}{h\nu_B} , \qquad (15.4)$$

which gives us

$$\nu_A \approx \left(1 - \frac{gH}{c^2}\right) \nu_B . \qquad (15.5)$$

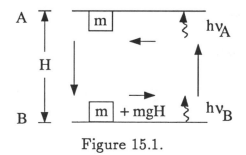

Figure 15.1.

d) We can think of the time difference between the cosmonaut's clock and the earthbound clock as being due to two separate effects. The first effect is the simple special-relativistic time dilation due to the differing velocities of the two clocks. According to Newtonian mechanics, the velocity of the cosmonaut is $v \approx \sqrt{R_E g}$, where we approximate the cosmonaut's orbital radius as the radius of the earth R_E. (The cosmonaut might be on the order of 100 miles above the earth's surface, while $R_E \approx 4000$ miles.) In comparison, the velocity of the earth observer's clock is negligible, and the ratio of the two times is

$$\frac{\Delta t_C}{\Delta t_E} = \sqrt{1 - \frac{v^2}{c^2}} \approx 1 - \frac{gR_E}{2c^2} , \qquad (15.6)$$

where Δt_C is the time measured by the cosmonaut, and Δt_E is the time measured on earth.

The second effect is identical to the gravitational redshift discussed in part (c). Because the two clocks are at different heights, they will run at different speeds. In fact, by (15.5) the ratio is

$$\frac{\Delta t_C}{\Delta t_E} \approx 1 - \frac{gH}{c^2} , \qquad (15.7)$$

where H is the height of the cosmonaut above the earth. For a low earth orbit, this effect is completely negligible compared to the time dilation (15.6).

If we wished, we could have solved this problem in the Schwarzchild metric, and automatically obtained both effects.

e) Define the time between explosions to be Δt_E as measured on the earth and Δt_0 as measured in the distant galaxy. The time between explosions is the inverse of a frequency, which will redshift just like the frequency of a photon. The redshift z is defined such that

$$\frac{\nu_0}{\nu_E} = 1 + z , \qquad (15.8)$$

so

$$\frac{\Delta t_E}{\Delta t_0} = 1 + z . \qquad (15.9)$$

Solution 6.2. a) We will find x', the distance from the rocket to the earth's center at proper time τ, as measured in the rocket's frame. Then the angle subtended by the earth can be found using $\tan(\theta'/2) = R_e/x'$, where R_e is the earth's radius.

Let t be the elapsed time and x the distance from the center of the earth to the rocket, in the reference frame of an observer on the earth. Then the four-velocity of the rocket in the earth's frame is

$$U = \left(\frac{dt}{d\tau}, \frac{dx}{d\tau}\right) = \gamma(1, \mathbf{v}), \tag{15.10}$$

where $\mathbf{v} = v\hat{\mathbf{x}}$ is the rocket's velocity measured in the earth's frame. The four-acceleration is $a \equiv dU/d\tau$, with

$$a \cdot a = -\left(\frac{d^2t}{d\tau^2}\right)^2 + \left(\frac{d^2x}{d\tau^2}\right)^2. \tag{15.11}$$

Now if we consider an inertial frame, with coordinates $\tilde{t}$ and $\tilde{x}$, instantaneously comoving with the rocket, we have that

$$\frac{d^2\tilde{t}}{d\tau^2} = 0 \quad \text{and} \quad \frac{d^2\tilde{x}}{d\tau^2} = \frac{d^2\tilde{x}}{d\tilde{t}^2} = g, \tag{15.12}$$

so that in the comoving frame,

$$a \cdot a = g^2. \tag{15.13}$$

Since this is an invariant quantity, it has this value in any frame.

Because $U \cdot U = -1$, we can take a derivative to get $U \cdot a = 0$. Using this and equation (15.10), we have

$$\left(\frac{d^2t}{d\tau^2}\right) = \left(\frac{d^2x}{d\tau^2}\right) v. \tag{15.14}$$

Combining equations (15.11), (15.13), and (15.14) gives

$$(1 - v^2)\left(\frac{d^2x}{d\tau^2}\right)^2 - g^2, \tag{15.15}$$

which we can rearrange to give $d^2x/d\tau^2 = \gamma g$. We can write

$$\frac{d^2x}{d\tau^2} = \frac{d}{d\tau}(\gamma v) = \gamma^3\frac{dv}{d\tau}, \tag{15.16}$$

or

$$\frac{dv}{d\tau} = g(1 - v^2).$$ (15.17)

Integrating this equation and using the boundary condition that $v = 0$ at $\tau = 0$ gives $v = \tanh g\tau$, $\gamma = \cosh g\tau$, and $dx/d\tau = \gamma v = \sinh g\tau$. Then the distance to the rocket in the earth's frame is

$$x = \int_0^\tau \sinh g\tau \, d\tau = \frac{1}{g} \cosh g\tau - x(0) = \frac{1}{g} \cosh g\tau - R_e.$$ (15.18)

The transformation to the rocket's frame is a simple length contraction, so the distance x' that the rocket measures is

$$x' = \frac{x}{\gamma} = \frac{\cosh g\tau - gR_e}{g \cosh g\tau}.$$ (15.19)

The radius of the earth perpendicular to the rocket's direction of travel is unaffected by the transformation to the rocket's frame, so if θ' is the angular size of the earth as seen by the rocket then

$$\tan \frac{\theta'}{2} = \frac{R_e}{x'} = \frac{R_e g \cosh g\tau}{\cosh g\tau - gR_e}.$$ (15.20)

b) As τ gets large, we have $\tan \theta'/2 \to gR_e$. As long as $gR_e \ll 1$ (which is true as long as $g \ll 10^7$ km/s^2!), we can use the small angle approximation and write $\theta' \approx 2gR_e$.

Solution 6.3. We use geometrized units in this problem: $G = c = 1$. Since the cylinder mass is small (i.e., $\rho R^2 \ll 1$), we can use linearized theory, in which the metric is written as $g_{\alpha\beta} = \eta_{\alpha\beta} + h_{\alpha\beta}$, where $\eta_{\alpha\beta}$ is the metric for flat space in cylindrical coordinates and $h_{\alpha\beta} \ll \eta_{\alpha\beta}$. Light travels along a null trajectory, so that $ds^2 = g_{\alpha\beta}dx^\alpha dx^\beta = 0$. Thus, in order to find Δt to lowest order we must find $h_{\alpha\beta}$ (and thus $g_{\alpha\beta}$) to first order in the small quantities w and ρR^2.

The result that we need from linearized theory is

$$\Box \bar{h}^{\alpha\beta} = \eta^{\mu\nu} \bar{h}^{\alpha\beta}_{,\mu\nu} = -16\pi T^{\alpha\beta},$$ (15.21)

where $T^{\alpha\beta}$ is the stress-energy tensor and

$$h^{\alpha\beta} = \overline{h}^{\alpha\beta} - \frac{1}{2}\eta^{\alpha\beta}\overline{h}. \tag{15.22}$$

(Equation (15.21) is true in the Lorentz gauge, $\overline{h}^{\mu\nu}{}_{,\nu} = 0$. For an excellent derivation, see Schutz.)

First, we find an expression for Δt in terms of $h_{\alpha\beta}$, and we will see that we only need to compute one component of $h_{\alpha\beta}$. Note that we will find $h_{\alpha\beta}$ inside the cylinder, and since the metric must be continuous, it will be legitimate to extend $g_{\alpha\beta} = \eta_{\alpha\beta} + h_{\alpha\beta}$ to just outside the cylinder, where the light travels. We set $ds^2 = 0$ along the path of interest, where $dr = dz = 0$, and find an expression for dt^2 in terms of $d\phi^2$:

$$ds^2 = 0 = (-1 + h_{tt})dt^2 + 2h_{t\phi}dtd\phi + (r^2 + h_{\phi\phi})d\phi^2, \tag{15.23}$$

so that

$$dt^2 = \left(\frac{2h_{t\phi}(dt/d\phi) + r^2 + h_{\phi\phi}}{1 - h_{tt}}\right)d\phi^2. \tag{15.24}$$

We take the square root and expand it, letting $r \approx R$:

$$\sqrt{dt^2} \approx R\left(1 + \frac{h_{t\phi}}{R^2}\frac{dt}{d\phi} + \frac{h_{\phi\phi}}{2R^2}\right)\left(1 + \frac{h_{tt}}{2}\right)d\phi$$

$$\approx R\left(1 + \frac{h_{t\phi}}{R^2}\frac{dt}{d\phi} + \frac{h_{\phi\phi}}{2R^2} + \frac{h_{tt}}{2}\right)d\phi. \tag{15.25}$$

We will solve (15.25) iteratively. To zeroth order in $h_{\alpha\beta}$, light traveling clockwise around the cylinder has $dt/d\phi \approx R$, and for a counterclockwise path, $dt/d\phi \approx -R$. (That is, the direction of travel determines the sign of $dt/d\phi$.) To first order, inserting $(dt/d\phi) = \pm R$ on the right hand side of (15.25), we see that the only term contributing to Δt is the term proportional to $dt/d\phi$, so we have

$$\Delta t = 2\int_0^\pi h_{t\phi}d\phi. \tag{15.26}$$

Now we solve for $h_{t\phi}$. $T^{\alpha\beta}$ is given by

$$T^{\alpha\beta} = (\rho + p)U^\alpha U^\beta + pg^{\alpha\beta}. \tag{15.27}$$

Since the particles comprising the cylinder are very nonrelativistic, $p \approx 0$. The rotation is also nonrelativistic, so that $\gamma \approx 1$, and $U \approx (1, 0, U^\phi, 0)$, where $U^\phi = \omega$. Thus, the nonzero components of $T^{\alpha\beta}$ inside the cylinder are

$$
\begin{aligned}
T^{tt} &= \rho, \quad T^{\phi\phi} = \rho\omega^2 \approx 0, \\
T^{t\phi} &= \rho\omega, \quad T^{\phi t} = \rho\omega.
\end{aligned}
\tag{15.28}
$$

We are trying to find an answer to lowest order in ω, so we drop $\mathcal{O}(\omega^2)$.

Consider the differential equation for $\bar{h}^{\alpha\beta}$, (15.21). Because we are considering a steady-state situation, the time derivative vanishes, and if we assume that the cylinder's length is much greater than R, the derivatives with respect to z and ϕ also vanish. We are left with

$$
\nabla_r^2 \bar{h}^{\alpha\beta} \equiv \frac{1}{r}\frac{\partial}{\partial r}\left(r\frac{\partial}{\partial r}\bar{h}^{\alpha\beta}\right) = -16\pi T^{\alpha\beta}.
\tag{15.29}
$$

Since $T^{\alpha\beta}$ is constant, we can integrate directly to find

$$
\bar{h}^{\alpha\beta} = -4\pi T^{\alpha\beta} r^2 + A^{\alpha\beta}.
\tag{15.30}
$$

(We have dropped a term singular at the origin.) Here, $A^{\alpha\beta}$ is a constant, as yet undetermined.

Having found an expression for $\bar{h}^{\alpha\beta}$, we must now extract $h_{t\phi}$. Since $\bar{h}^{\alpha\beta}$ is already first-order, we may use the flat-space metric $\eta_{\alpha\beta}$ to lower its indices. Recall that in cylindrical coordinates, the nonzero components of $\eta_{\alpha\beta}$ are:

$$
\begin{aligned}
\eta_{tt} &= -1, \quad \eta_{rr} = 1, \\
\eta_{\phi\phi} &= r^2, \quad \eta_{zz} = 1.
\end{aligned}
\tag{15.31}
$$

Then, using the definition of $\bar{h}^{\alpha\beta}$, equation (15.22), we have

$$
h_{t\phi} = \eta_{tt}\eta_{\phi\phi}h^{t\phi} = \eta_{tt}\eta_{\phi\phi}\bar{h}^{t\phi} = 4\pi\rho\omega r^4.
\tag{15.32}
$$

For the last equality, we have set the constant $A_{t\phi}$ equal to zero because $h_{t\phi}$ must equal zero for $r = 0$, since the line element must be independent of ϕ at the origin. Using this $h_{t\phi}$ in the expression for Δt found above, equation (15.26), we find

$$
\Delta t = 2[4\pi\rho\omega R^4]\pi = 8\pi^2\rho\omega R^4.
\tag{15.33}
$$

We must multiply by G/c^4 to get units of time, yielding

$$\Delta t = \frac{8\pi^2 G}{c^4} \rho \omega R^4 \tag{15.34}$$

as the first-order time difference.

To estimate the magnitude of this Δt, we guess a density ρ of 10 g/cm^3, an angular velocity of $\omega \approx 1$ rad/s and $R \approx 10$ m (which requires a rather cavernous room). These values yield $\Delta t \approx 10^{-34}$ s, a very small time interval indeed.

The wavelength of green light is 540 nm so that its period is $T_{gr} = \lambda/c \approx 2 \times 10^{-15}$ s. Thus, the ratio of Δt to the period of optical light is approximately $\Delta t / T_{gr} \approx 10^{-19}$, which makes this experiment extremely impractical, albeit nifty.

———

Solution 6.4. a) The stress-energy tensor for a perfect fluid is

$$T^\alpha_{\ \beta} = \begin{pmatrix} -\rho & & & \\ & p & & \\ & & p & \\ & & & p \end{pmatrix}. \tag{15.35}$$

For a photon gas, which is a perfect fluid, the equation of state is $p = \rho/3$, and the stress-energy tensor is traceless. Einstein's equation is

$$8\pi T_{\alpha\beta} = R_{\alpha\beta} - \frac{1}{2} g_{\alpha\beta} \mathcal{R}, \tag{15.36}$$

where we have written the Ricci scalar as $R^\alpha_{\ \alpha} = \mathcal{R}$ to avoid confusion with the expansion factor R. Taking the trace of Einstein's equation and using the fact that $T^\alpha_{\ \alpha} = 0$, we find that the Ricci scalar vanishes. Then the tt component of Einstein's equation gives the relation

$$\frac{\ddot{R}}{R} + \frac{8\pi}{3} \rho = 0, \tag{15.37}$$

and the ii component gives

$$\frac{\ddot{R}}{R} + \frac{2\dot{R}^2}{R^2} + \frac{2k}{R^2} = \frac{8\pi}{3}\rho,\tag{15.38}$$

where we have used the components of the Ricci tensor as given in (6.2). Eliminating the density between equations (15.37) and (15.38) and substituting $k = 1$ for a closed universe, we can write a differential equation for the expansion factor R:

$$\frac{R\ddot{R}}{\dot{R}^2} + \frac{1}{\dot{R}^2} + 1 = 0.\tag{15.39}$$

If we evaluate the expression at time t_0, and rewrite it in terms of the deceleration parameter q_0 and the Hubble constant H_0, both of which are defined in the question, we find

$$R_0 = \sqrt{\frac{1}{(q_0 - 1)H_0^2}}.\tag{15.40}$$

Similarly, we can evaluate (15.37) at t_0 and use the relation $\rho_0 = aT_0^4$ to find the temperature at time t_0:

$$T_0 = \left(\frac{3q_0 H_0^2}{8\pi a}\right)^{1/4}.\tag{15.41}$$

b) Multiplying equation (15.39) by $\dot{R}^2$ and integrating once, we find

$$R\dot{R} + t = b,\tag{15.42}$$

and integrating again gives

$$R^2/2 + t^2/2 = bt + d,\tag{15.43}$$

where b and d are constants of integration. The boundary condition that $R = 0$ at $t = 0$ requires $d = 0$. Eliminating t between the two equations (15.42) and (15.43) then gives the constant b,

$$b = R(1 + \dot{R}^2)^{1/2}.\tag{15.44}$$

(Note that only the positive root can satisfy equation (15.42), since R, $\dot{R}$, and t are all positive quantities at early times.) Evaluating (15.44) at time t_0 and using our previous expression for R_0, (15.40), allows us to find $b = \sqrt{q_0}/[H_0(q_0 - 1)]$. So, solving equation (15.42) for t_0 gives

$$t_0 = \frac{1}{H_0(\sqrt{q_0} + 1)} . \tag{15.45}$$

To find $R(t)$ for small t, we drop the t^2 term in (15.43), which leaves

$$R(t) \approx \left(\frac{2\sqrt{q_0}}{H_0(q_0 - 1)} \right)^{1/2} t^{1/2} . \tag{15.46}$$

To find $T(t)$, take equations (15.37) and (15.38) and eliminate the term involving $\ddot{R}$. Substituting $\rho = aT^4$ gives

$$\frac{\dot{R}^2 + 1}{R^2} = \frac{8\pi a}{3} T^4 . \tag{15.47}$$

Multiplying through by R^4 and using equation (15.44) gives us that

$$(RT)^4 = \frac{3b^2}{8\pi a} . \tag{15.48}$$

In other words, for a radiation-dominated universe, RT is a constant (which is true for any value of k).

For early times, we can use expression (15.46) to find the temperature:

$$T(t) \approx \left(\frac{3}{32\pi a} \right)^{1/4} t^{-1/2} . \tag{15.49}$$

c) First we want to show that $R(t)$ really is an "expansion factor." Consider a particle at coordinate rest, say at $u = (u^t, 0, 0, 0)$. Then the equation of motion for the particle is

$$\frac{du^r}{d\tau} = -\Gamma^r_{\mu\nu} u^\mu u^\nu = -\Gamma^r_{tt} (u^t)^2 . \tag{15.50}$$

It is straightforward to calculate that $\Gamma^r_{tt} = 0$ in the Robertson-Walker metric, so a particle at coordinate rest *stays* at coordinate rest. Since the proper distance between two particles is the product of their coordinate separation and the scale factor $R(t)$, the proper size of a physical region scales as $R(t)$.

Applying this reasoning, we can see that a region of size L_0 at time t_0 will have evolved from a region of size $L_1 = L_0(R_1/R_0)$ at time t_1. If we assume that both t_0 and t_1 are small enough that the approximation (15.46) is valid, then we can use (15.45) to find that L_1 is given by

$$L_1 \approx (t_0 t_1)^{1/2} = \left(\frac{t_1}{H_0(\sqrt{q_0} + 1)} \right)^{1/2}. \tag{15.51}$$

The greatest distance between two points which are causally connected at time t is t itself, so the number of causally-connected regions at time t_1 is

$$N \approx \left(\frac{L_1}{t_1} \right)^3 = \left(\frac{1}{H_0(\sqrt{q_0} + 1)t_1} \right)^{3/2}. \tag{15.52}$$

Notice that this expression is for a radiation-dominated universe. If the universe is matter-dominated, R varies as $t^{2/3}$, rather than $t^{1/2}$ (e.g., see Kolb and Turner). However, since the universe did most of its expansion while it was still radiation-dominated, (15.52) is a reasonable approximation if we take t_1 to be sufficiently early.

To get a numerical estimate, we combine equation (15.52) with the expression for the temperature (15.49) to eliminate the unknown time t_1:

$$N \approx \left(\frac{32\pi^3 (k_B T)^4}{45\hbar^3 H_0^2 (\sqrt{q_0} + 1)^2} \right)^{3/4}. \tag{15.53}$$

Using units where $c = G = 1$, we have 1 GeV$= 1.3 \times 10^{-54}$ m. It is simplest to convert all dimensional quantities to the same unit, say meters. Then $\hbar = 2.6 \times 10^{-70} \text{m}^2$, $k_B T = 1.3 \times 10^{-39} \text{m}$, and $H_0 = 1.1 \times 10^{-26} \text{m}^{-1}$. Using these values and $q_0 = 2$, we find $N \sim 10^{79}$.

d) The homogeneity of the visible universe is surprising because there are no physical processes currently acting which can smooth out inhomogeneities over distances as great as the horizon length. If we extrapolate back to a time when such processes *were* taking place (for sake of argument we take this to be the GUT time t_1), then we find that the visible universe has evolved from 10^{79} regions that were causally disconnected at t_1, and we would not expect any correlation between these different regions.

The theory of inflation proposes that at some point in the early history of the universe (again we choose a time of order t_1), there was a huge expansion in a very short period, so that the expansion factor $R(t)$ increased by many orders of magnitude, while the temperature before and after inflation was almost the same. One possible mechanism for this is that in the very early universe, the vacuum had a nonzero energy density Λ due to some scalar field which played a symmetry-breaking role similar to that of the electroweak Higgs field. Thus there was an extra contribution to the stress-energy tensor (15.35), the Ricci scalar $\mathcal{R}$ did not vanish, and the equations of motion (15.37) and (15.38) were modified correspondingly. For early enough times, the photon energy density dominated, but as $R(t)$ increased the photons were red-shifted, while Λ stayed constant, until the two contributions were comparable. At this point the equation governing $R(t)$ became $\ddot{R}/R \approx$ constant, which means that R grew exponentially in time. The remaining photons got red-shifted away, and the universe supercooled. However this did not go on for ever, as the universe eventually underwent a phase transition to the true vacuum, and inflation stopped. In the process, the vacuum energy was converted to scalar particles, which then thermalized back into photons with energy density approximately Λ. Consequently the temperature after inflation was nearly the same as that before, while the scale factor had increased by a huge factor.

As a consequence of all this, a small, causally-connected, smooth region of space could be inflated to a volume about 10^{79} times greater, so that different points in that region would lie outside the horizon, but still be correlated. During the subsequent expansion of the universe, more of this smooth region would come back within the horizon, and the universe would appear homogeneous.

For a radiation-dominated universe, the entropy in a physical volume V is

$$S = \int dV \frac{4}{3} a T^3 \propto R^3 T^3. \tag{15.54}$$

We saw in part (b), equation (15.48), that the product RT is constant, and so during the era of radiation domination, the expansion is adiabatic.

However, we argued that during inflation, R^3 increased by a factor of about 10^{79}, while T stayed constant. Thus the relative increase in entropy during this process is at least 10^{79}. (For more details on inflation, a good reference is Kolb and Turner.)

Solution 6.5.   a) The particle travels along a path described by the geodesic equation,

$$m\frac{dp_\beta}{d\tau} = \frac{1}{2}g_{\nu\alpha,\beta}p^\nu p^\alpha .$$ (15.55)

Since $g_{\nu\alpha}$ is independent of t and ϕ, we can immediately say that p_t and p_ϕ are constants of the motion. Define the conserved quantities $E \equiv -p_t/m$ and $L \equiv p_\phi/m$. If we take the motion to be in the $\theta = \pi/2$ plane, then by (15.55) $p_\theta = 0$ and we can write

$$p \cdot p = -m^2 = g_{tt}p^t p^t + g_{\phi\phi}p^\phi p^\phi + g_{rr}p^r p^r .$$ (15.56)

We insert $p^r = m(dr/d\tau)$, $p^t = g^{tt}p_t = p_t/g_{tt}$, and $p^\phi = g^{\phi\phi}p_\phi = p_\phi/g_{\phi\phi}$ into expression (15.56), and then solve for $(dr/d\tau)$:

$$\left(\frac{dr}{d\tau}\right)^2 = E^2 - \left(1 - \frac{2M}{R}\right)\left(1 + \frac{L^2}{R^2}\right)$$ (15.57)

$$\equiv E^2 - V^2 ,$$

which is the desired "energy" conservation equation for radial motion.

To find the relationship between p_ϕ, M and R, we differentiate with respect to proper time,

$$\frac{d}{d\tau}\left(\frac{dr}{d\tau}\right)^2 = 2\frac{dr}{d\tau}\frac{d^2r}{d\tau^2} = -\frac{dV^2}{d\tau} = -\frac{dr}{d\tau}\frac{dV^2}{dr} .$$ (15.58)

This relation holds for a general, non-circular orbit. We cancel $dr/d\tau$ on both sides, and then note that for a circular orbit $d^2r/d\tau^2 = 0$, so we have $dV^2/dr = 0$. Taking this derivative leads to the condition that

$$L^2 = \left(\frac{p_\phi}{m}\right)^2 = \frac{MR^2}{R - 3M} .$$ (15.59)

Solving for R in (15.59) gives two roots corresponding to two possible circular orbits, only one of which is stable. The stable root is

$$R = \frac{L^2}{2M}\left(1 + \sqrt{1 - \frac{12M^2}{L^2}}\right) . \tag{15.60}$$

b) To find the orbital period as measured by a clock carried along with the particle, we use the relation

$$\frac{d\phi}{d\tau} = \frac{p^\phi}{m} = g^{\phi\phi}\frac{p_\phi}{m} = \frac{1}{R^2}\sqrt{\frac{MR^2}{R - 3M}} . \tag{15.61}$$

Integrating with respect to τ around one orbit gives the period,

$$\Delta\tau = 2\pi R\sqrt{\frac{R}{M} - 3} . \tag{15.62}$$

c) The far away observer measures the period of the orbit in coordinate time, since for that observer $2M/R \ll 1$ and $g_{tt} = 1$. Thus we need to convert our previous result from the particle's proper time to coordinate time. To do this, we note that

$$\begin{aligned}
\frac{d\phi}{dt} &= \frac{d\phi}{d\tau}\frac{d\tau}{dt} = \frac{p^\phi}{m}\frac{m}{p^t} = \frac{g^{\phi\phi}}{g^{tt}}\frac{p_\phi}{p_t} , \\
&= \frac{(1 - 2M/R)}{R^2}\left(\frac{L}{E}\right) .
\end{aligned} \tag{15.63}$$

The orbit is circular, so $dr/d\tau = 0$, and we can use the energy conservation equation (15.57) to show that

$$E^2 = V^2 = \frac{(1 - 2M/R)^2}{(1 - 3M/R)} , \tag{15.64}$$

where we have used the value for L found in part (a), (15.59). We then find that

$$\frac{L}{E} = \frac{\sqrt{MR}}{(1 - 2M/R)} . \tag{15.65}$$

Substituting this into (15.63) gives

$$\frac{d\phi}{dt} = \frac{\sqrt{MR}}{R^2} , \tag{15.66}$$

and thus the distant observer measures a period

$$\Delta t = 2\pi R \sqrt{\frac{R}{M}} . \tag{15.67}$$

d) A stationary observer at radius R has a proper time interval given by

$$d\tau'^2 = \left(1 - \frac{2M}{R}\right) dt^2 , \tag{15.68}$$

since $dr = d\theta = d\phi = 0$. It follows that

$$\frac{d\phi}{d\tau'} = \frac{d\phi}{dt}\frac{dt}{d\tau'} = \frac{d\phi}{dt}\left(1 - \frac{2M}{R}\right)^{-1/2} , \tag{15.69}$$

so that the oberver at fixed R measures an orbital period of

$$\Delta\tau' = 2\pi R \sqrt{\frac{R}{M} - 2} . \tag{15.70}$$

Solution 6.6.   a) The acceleration four-vector a^μ of a particle in a curved spacetime is defined as the covariant convective derivative of the velocity four-vector U,

$$a^\mu = (U^\alpha \nabla_\alpha U)^\mu = \frac{\partial U^\mu}{\partial \tau} + \Gamma^\mu{}_{\alpha\beta} U^\alpha U^\beta . \tag{15.71}$$

Thus, even if the particle is at coordinate rest, with $U = (U^t, 0, 0, 0)$, the Christoffel symbols can give rise to a nonzero acceleration.

Now consider a rocket in the gravitational field of a nearby star, where the spacetime curvature is given by the Schwarzschild metric,

$$ds^2 = -\left(1 - \frac{2M}{r}\right)dt^2 + \left(1 - \frac{2M}{r}\right)^{-1}dr^2 + r^2 d\Omega^2. \qquad (15.72)$$

The rocket is holding itself at fixed radius and constant angle: consequently its velocity vector has just one nonvanishing component, U^t. Since the four-velocity of a massive particle always obeys the equation $U \cdot U = -1$, we can find the value of this component:

$$-1 = -\left(1 - \frac{2M}{R}\right)\left(U^t\right)^2, \qquad (15.73)$$

or

$$U^t = \left(1 - \frac{2M}{R}\right)^{-1/2}. \qquad (15.74)$$

The expression for the acceleration also simplifies to $a^\mu = \Gamma^\mu{}_{tt} U^t U^t$. It is easy to see that only one of the acceleration components does not vanish, namely a^r, and we have to evaluate a single Christoffel coefficient,

$$\Gamma^r{}_{tt} = \frac{1}{2}g^{r\beta}\left(g_{\beta t,t} + g_{t\beta,t} - g_{tt,\beta}\right) = -\frac{1}{2}g^{rr}g_{tt,r}. \qquad (15.75)$$

After a small amount of algebra we find that this nonzero component of acceleration is $a^r = M/R^2$.

The invariant acceleration is

$$\alpha = \sqrt{a^\beta a_\beta} = \sqrt{g_{rr}a^r a^r} = \frac{M}{R^2\sqrt{1 - 2M/R}}. \qquad (15.76)$$

b) For simplicity we will now work in the frame of an observer $\mathcal{O}$ who is dropped from the rocket, so that at the rocket's proper time τ_0, $\mathcal{O}$ is instantaneously *at rest* with respect to the rocket. Note that the observer is falling freely, and hence in $\mathcal{O}$'s coordinate system spacetime is locally flat and we can use special relativity.

$\mathcal{O}$ will observe the rocket accelerating outwards, and will measure the magnitude of the acceleration to be simply α. We know this because

α is an invariant quantity, and the other components of the acceleration still vanish in $\mathcal{O}$'s frame: in particular, we can use the identity $U \cdot a = 0$ and the instantaneous value of the four-velocity, $U = (1, 0, 0, 0)$, to see that $a^t = 0$. Thus $\alpha = \sqrt{g_{rr} a^r a^r}$, as before. $\mathcal{O}$ can now calculate the force required to produce such an acceleration, namely $F = m(\tau_0)\alpha$, where $m(\tau_0)$ is the rest mass of the rocket at proper time τ_0. The rocket motor produces this force by ejecting a beam of photons radially inwards towards the star. In time $\delta\tau$, the momentum of the photons ejected must equal $m(\tau_0)\alpha\,\delta\tau$, and if we work in units with $c = 1$ then this will also equal the energy of the ejected photons. By conservation of energy, the energy of the rocket (or equivalently its mass) must decrease by the same amount. In this way we see that $m(\tau)$ obeys the differential equation

$$\frac{dm(\tau)}{d\tau} = -m(\tau)\alpha, \tag{15.77}$$

which we can solve to find

$$m(\tau) = m(0)e^{-\alpha\tau}. \tag{15.78}$$

c) Suppose now the rocket is orbiting at radius $r = R$ in the plane where $\theta = \pi/2$, with proper angular frequency $d\phi/d\tau = \omega$. The velocity four-vector now has two nonzero components: U^t and $U^\phi \equiv \omega$. Once again we use the condition $U \cdot U = -1$ to find

$$\left(U^t\right)^2 = \frac{1 + R^2\omega^2}{1 - 2M/R}. \tag{15.79}$$

If we evaluate the four-acceleration (15.71), we find

$$a^\mu = \Gamma^\mu{}_{tt} U^t U^t + 2\Gamma^\mu{}_{t\phi} U^t U^\phi + \Gamma^\mu{}_{\phi\phi} U^\phi U^\phi. \tag{15.80}$$

In fact, all components of $\Gamma^\mu{}_{\phi t}$ vanish, and we find that the acceleration is still only in the radial direction,

$$\begin{aligned}
a^r &= -\frac{1}{2}g^{rr}g_{tt,r} U^t U^t - \frac{1}{2}g^{rr}g_{\phi\phi,r} U^\phi U^\phi \\
&= \frac{M}{R^2}(1 + R^2\omega^2) - \left(1 - \frac{2M}{R}\right)R\omega^2 \\
&= \frac{M}{R^2} + 3M\omega^2 - R\omega^2.
\end{aligned} \tag{15.81}$$

The magnitude of the invariant acceleration is thus

$$\alpha^2 = \left(\frac{M}{R^2} + (3M - R)\omega^2\right)^2 \left(1 - \frac{2M}{R}\right)^{-1}. \qquad (15.82)$$

Therefore α vanishes if

$$\omega^2 = \frac{M}{R^2(R - 3M)}. \qquad (15.83)$$

This is the familiar result for a particle in a free orbit around a black hole.

Solution 6.7. a) The component R^x_{zxz} of the Riemann tensor is given by

$$R^x_{zxz} = \Gamma^x_{zz,x} - \Gamma^x_{xz,z} + \Gamma^x_{xf}\Gamma^f_{zz} - \Gamma^x_{zf}\Gamma^f_{xz}, \qquad (15.84)$$

where the Christoffel symbols are given in turn by

$$\Gamma^\alpha_{\beta\delta} = \frac{1}{2}g^{\alpha\lambda}(g_{\lambda\beta,\delta} + g_{\lambda\delta,\beta} - g_{\beta\delta,\lambda}). \qquad (15.85)$$

Let us consider each term on the right hand side of equation (15.84) separately. Since the metric has no dependence on x or y, only derivatives with respect to z or t survive and the only nonzero terms of (15.84) are:

$$\Gamma^x_{xz,z} = \frac{1}{2}(g^{xx}g_{xx,z})_{,z}, \qquad (15.86)$$

and

$$\Gamma^x_{zf}\Gamma^f_{xz} = (\Gamma^x_{xz})^2. \qquad (15.87)$$

Therefore the equation for R^x_{zxz}, (15.84), reduces to

$$R^x_{zxz} = -\frac{1}{2}(g^{xx}g_{xx,z})_{,z} - \frac{1}{4}(g^{xx}g_{xx,z})^2. \qquad (15.88)$$

To evaluate this expression, we will assume that we are in the $z > 0$ region of space so that

$$g_{xx} = (1 - kz)^2 e^{2kt}, \tag{15.89}$$
$$g^{xx} = (1 - kz)^{-2} e^{-2kt}, \tag{15.90}$$
$$g_{xx,z} = -2k(1 - kz)e^{2kt}. \tag{15.91}$$

Using these values in the last equation for R^x_{zxz}, (15.88), yields the desired result that $R^x_{zxz} = 0$. (Of course, we get zero for $z < 0$ as well.)

b) A particle follows a geodesic if its velocity four-vector U^α obeys

$$U^\alpha U^\beta_{;\alpha} = \frac{dU^\beta}{\tau} + U^\alpha U^\nu \Gamma^\beta_{\alpha\nu} = 0. \tag{15.92}$$

We wish to find the acceleration $a^\alpha = dU^\alpha/d\tau$. Assume the particle is at an infinitesimal distance ϵ from the wall and is initially at rest, so that $U = (1, 0, 0, 0)$. (We will assume that $\epsilon > 0$ without loss of generality.) Under these initial conditions, the geodesic equation (15.92) becomes

$$\frac{dU^\beta}{d\tau} = -\Gamma^\beta_{tt}. \tag{15.93}$$

Now, it is easily seen that $\Gamma^\beta_{tt} \neq 0$ only for $\beta = z$. Therefore

$$a^z = \frac{dU^z}{d\tau} = -\Gamma^z_{tt} = k(1 - k\epsilon). \tag{15.94}$$

We see that for $\epsilon < 1/k$, we have $d^2z/d\tau^2 > 0$, and the particle accelerates away from the wall. The instantaneous acceleration four-vector is $a = (0, 0, 0, k(1 - k\epsilon))$. The invariant acceleration is

$$\sqrt{a^\alpha a_\alpha} = k(1 - k\epsilon) \approx k \tag{15.95}$$

for $\epsilon \ll 1/k$.

_____

Solution 6.8.   a) Since S evolves by parallel transport, its equation of motion is given by $U^\alpha D^\beta_{;\alpha} = 0$, or

$$\frac{dS^\beta}{d\tau} = -U^\alpha \Gamma^\beta_{\nu\alpha} S^\nu. \qquad (15.96)$$

We make the natural choice of coordinates such that the motion is confined to the equatorial plane, i.e., $\theta = \pi/2$. With this choice of coordinates, a circular geodesic orbit in the Schwarzschild metric has a four-velocity U whose components are $U^r = U^\theta = 0$, $U_t = -E$, and $U_\phi = L$ where E and L are constants (see Problem 6.5). The equation of motion (15.96) reduces to

$$\frac{dS^\beta}{d\tau} = -U^t \Gamma^\beta_{\nu t} S^\nu - U^\phi \Gamma^\beta_{\nu\phi} S^\nu. \qquad (15.97)$$

Next we use the additional result from stable circular motion in a Schwarzschild metric that

$$\omega \equiv \frac{d\phi}{dt} = \frac{U^\phi}{U^t} = \sqrt{\frac{M}{R^3}} \qquad (15.98)$$

to simplify the equation of motion even further. (Note that the result (15.98) is easily remembered since it happens to be identical in form to the corresponding Newtonian result.) We divide through by U^t and obtain

$$\omega \frac{dS^\beta}{d\phi} = -\Gamma^\beta_{\nu t} S^\nu - \omega \Gamma^\beta_{\nu\phi} S^\nu. \qquad (15.99)$$

Now we can find explicit differential equations for the evolution of the spatial components of S with respect to the variable ϕ. First let us consider the θ-component of (15.99). We calculate the Christoffel symbols in the usual fashion from the metric (and remember that $\theta = \pi/2$) to find $\Gamma^\theta_{\nu t} = \Gamma^\theta_{\nu\phi} = 0$, so that $dS^\theta/d\theta = 0$.

Next consider $\beta = \phi$ in (15.99), in which case the only nonvanishing relevant Christoffel symbol is $\Gamma^\phi_{r\phi}$. The differential equation governing the behavior of S^ϕ is

$$\frac{dS^\phi}{d\phi} = -\Gamma^\phi_{\phi r} S^r = -\frac{1}{R} S^r. \qquad (15.100)$$

Finally if $\beta = r$, the relevant nonzero Christoffel symbols are Γ^r_{tt} and $\Gamma^r_{\phi\phi}$, so that

$$\frac{dS^r}{d\phi} = -\frac{1}{\omega}\Gamma^r_{tt}S^t - \Gamma^r_{\phi\phi}S^\phi. \qquad (15.101)$$

Now we use the fact that $S \cdot U = 0$ and solve for S^t:

$$S^t = -\omega\frac{g_{\phi\phi}}{g_{tt}}S^\phi . \qquad (15.102)$$

We can use this to eliminate S^t from (15.101). Evaluating the components of the metric and their derivatives along the orbit, we find

$$\frac{dS^r}{d\phi} = (R - 3M)S^\phi. \qquad (15.103)$$

b) We can combine the coupled first-order differential equations (15.100) and (15.103) into two second-order differential equations,

$$\frac{d^2 S^\alpha}{d\phi^2} = -\left(1 - \frac{3M}{R}\right)S^\alpha, \qquad (15.104)$$

where $\alpha = \phi$ or r. These equations describe a spin precessing in ϕ-r space, rotating by an angle of $\sqrt{1 - 3M/R}\,\Delta\phi$ radians in the ϕ-r plane, as the gyroscope moves by an angle $\Delta\phi$ around its circular orbit. Suppose an observer at a fixed point on the orbit notes the direction that the spin is pointing every time the gyroscope passes her. For one orbit, ϕ changes by 2π, and the spin rotates by an angle $2\pi\sqrt{1 - 3M/R}$ in the ϕ-r plane. Therefore the spin will have rotated by an angle $2\pi(\sqrt{1 - 3M/R}-1)$ from its initial direction. From the fixed observer's point of view, the gyroscope has an orbital period of $(2\pi/\omega)\sqrt{1 - 2M/R}$ (see Problem 6.5). Therefore, the observer sees the gyroscope precessing at a rate

$$\frac{\sqrt{1 - 3M/R} - 1)\, 2\pi}{(\sqrt{1 - 2M/R})(2\pi/\omega)} \approx -\frac{3}{2}\sqrt{\frac{M^3}{R^5}}, \qquad (15.105)$$

if $M/R \ll 1$. If the gyroscope is not in a geodesic orbit then the equations of motion of the spin are not given by parallel transport

(15.96) but by the more general Fermi-Walker transport (see Lightman *et. al.*).

Solution 6.9. a) For a dust particle to remain in the solar system, it is necessary that the attractive gravitational force on it be greater than the repulsive force due to radiation pressure. At a distance r from the sun, the former has magnitude

$$F_G = \frac{GM_\odot m}{r^2}, \qquad (15.106)$$

where m is the mass of the particle, which we take to be a sphere of radius a and density ρ, so that $m = (4/3)\pi a^3 \rho$. We will choose its density to be $\rho \approx 1$ g/cm^3. The radiation force is approximately the radiation pressure times the geometric cross-section of the dust grain:

$$F_R \approx \pi a^2 \frac{L_\odot/c}{4\pi r^2}. \qquad (15.107)$$

We can now find the condition on the size of a particle for which $F_G > F_R$:

$$a > \frac{3L_\odot}{16\pi c G M_\odot \rho} \approx 6 \times 10^{-5} \text{ cm}, \qquad (15.108)$$

which is a very small dust grain.

b) The drag is a relativistic effect. In a frame co-moving with the orbiting dust particle, photons from the sun will appear to the particle to be coming from a slightly forward direction. The dust particle will absorb these photons and re-emit them isotropically in its rest frame, and will consequently absorb a small momentum flux in the direction opposite to its velocity.

Let us set $c = 1$ and restore it at the very end. We set up cartesian coordinates centered on the particle such that the sun lies on the $\hat{y}$-axis and the instantaneous motion $\mathbf{v}$ of the particle is in the x-direction. In the sun's rest frame, the photons incident on the dust particle have four-momenta given by $p = (E, 0, E, 0)$. If we boost to the particle's rest frame, the photons' four-momenta become $p' = (\gamma E, -\gamma v E, E, 0)$. When the particle scatters these photons, it absorbs (on average) the x-component of p'. Therefore, the particle feels a force opposite to $\hat{v}$ given by its geometrical cross-section times the flux of momentum in the x-direction, or

$$\mathbf{F}_d = -(\pi a^2)\left[\gamma \mathbf{v}\frac{L}{4\pi r^2}\right] \approx -F_R \mathbf{v}. \tag{15.109}$$

The torque resulting from this drag force causes the particle to spiral inward. If we assume the drag force is small, then the dust particle is in an almost circular orbit, whose radius is slowly decreasing. In plane-polar coordinates centered on the sun, the equations of motion for the particle are the radial force equation,

$$m\ddot{r} = F_R + m\omega^2 r - F_G \approx 0, \tag{15.110}$$

where $\omega = \dot{\theta}$, and the torque equation,

$$-rF_d = -rvF_R = \frac{d}{dt}(mr^2\omega). \tag{15.111}$$

If we define the constants $k = GM_\odot$ and $b = L_\odot \pi a^2/4\pi mc$, then equation (15.110) becomes

$$r^3\omega^2 = (k - b). \tag{15.112}$$

We can also rewrite equation (15.111) as

$$\frac{d}{dt}(r^2\omega) = -b\omega. \tag{15.113}$$

Solving for ω from equation (15.112) and eliminating it from equation (15.113) yields

$$\frac{dr}{dt} = -\frac{2b}{r}, \tag{15.114}$$

which we can integrate to find the total time T to fall:

$$T = -\frac{c}{2b}\int_{R_{\text{earth}}}^{R_\odot} r\, dr \approx \frac{R_{\text{earth}}^2 c}{4b}, \tag{15.115}$$

where we have reinstated c so that T has units of seconds. To find a numerical estimate for T, let us take a dust grain near the stability limit of part (a). For example, let $F_R = F_G/2$. Then $b = k/2 = GM_\odot/2$, so that $T \approx 2.5 \times 10^{10}$ s ≈ 1000 years.

Solution 6.10. a) In vacuum, Einstein's equations are

$$G_{\alpha\beta} \equiv R_{\alpha\beta} - \frac{1}{2} g_{\alpha\beta} R = 0. \qquad (15.116)$$

Here, $R_{\alpha\beta}$ is the Ricci tensor with trace R. In terms of the Riemann tensor, the $R_{\alpha\beta}$ is defined by $R_{\alpha\beta} = R^\mu_{\alpha\mu\beta}$.

We multiply through equation (15.116) with $g^{\alpha\beta}$ and find $R = 0$. In turn, putting $R = 0$ in equation (15.116) gives us $R_{\alpha\beta} = 0$, and from this will come conditions on α, β, and γ.

We must first calculate the Christoffel symbols from

$$\Gamma^\sigma_{\alpha\beta} = \frac{1}{2} g^{\sigma\delta} (g_{\delta\alpha,\beta} + g_{\delta\beta,\alpha} - g_{\alpha\beta,\delta}), \qquad (15.117)$$

which leads to the following nonzero terms:

$$\begin{aligned}
\Gamma^t_{xx} &= \alpha t^{2\alpha-1}, & \Gamma^t_{yy} &= \beta t^{2\beta-1}, & \Gamma^t_{zz} &= \gamma t^{2\gamma-1}, \\
\Gamma^x_{xt} &= \alpha t^{-1}, & \Gamma^y_{yt} &= \beta t^{-1}, & \Gamma^z_{zt} &= \gamma t^{-1}.
\end{aligned} \qquad (15.118)$$

We consider $R_{xx} = 0$, use the definition of the Riemann tensor, and find

$$\alpha(\alpha + \beta + \gamma - 1) = 0. \qquad (15.119)$$

We find similar equations, with $\alpha \leftrightarrow \beta$ or $\alpha \leftrightarrow \gamma$ from $R_{yy} = 0$ and $R_{zz} = 0$. Next we consider $R_{tt} = 0$ to find the final condition:

$$(\alpha + \beta + \gamma) - (\alpha^2 + \beta^2 + \gamma^2) = 0. \qquad (15.120)$$

b) A Killing vector is any four-vector field K that satisfies Killing's equation:

$$K_{\alpha;\beta} + K_{\beta;\alpha} = 0. \qquad (15.121)$$

Given such a vector, and any four-vector p that satisfies the geodesic equation of motion $\nabla_p p = 0$, the scalar quantity $K_\alpha p^\alpha$ is a constant along the geodesic. If we consider the case when p is the four-momentum of a freely-falling particle, then each conserved quantity of the motion is associated with a different K. These conserved quantities might include the energy, linear momentum, angular momentum, etc.

Now consider the Kasner metric. Since it is independent of x, y, and z, the quantities p_x, p_y, and p_z are conserved. (This is a standard result. See Schutz, Chapter 7.) We know that there are Killing vectors associated with each one, e.g., $p_x = p_\alpha K^\alpha$. In this way we write down three Killing vectors:

$$\begin{pmatrix} 0 \\ 1 \\ 0 \\ 0 \end{pmatrix}, \quad \begin{pmatrix} 0 \\ 0 \\ 1 \\ 0 \end{pmatrix}, \quad \text{and} \quad \begin{pmatrix} 0 \\ 0 \\ 0 \\ 1 \end{pmatrix}. \tag{15.122}$$

These correspond to the translational symmetries of the metric. As one might guess, it turns out that these three are the only independent Killing vectors for $\alpha \neq \beta$, $\alpha \neq \gamma$, $\beta \neq \gamma$. If $\alpha = \beta \neq \gamma$, we pick up the Killing vector describing rotation in the x-y plane (because if $\alpha = \beta$, then x and y are treated identically in the metric), and similarly $\alpha = \gamma$ leads to the vector describing rotations in the x-z plane, and $\gamma = \beta$, the vector describing rotation in the y-z plane. Of course, if $\alpha = \beta = \gamma = 0$, space is flat, and we pick up a Killing vector for time translation, and three vectors for "boosts," which look like rotations in the tx, ty, and tz planes. These vectors arise because t is a variable treated like x, y, and z in flat space. (Notice that $\alpha = \beta = \gamma \neq 0$ is not consistent with conditions (15.119) and (15.120).)

Although we have correctly guessed the solution above (except for one special case when $\alpha = 1$ and $\beta = \gamma = 0$), we will go through the somewhat laborious process of calculating the Killing vectors directly from the metric, both to demonstrate the technique and to confirm that we have found all the vectors.

Killing's equation (15.121) is really ten coupled equations. Letting $\epsilon = \delta \neq t$ leads to

$$K_{x,x} = \alpha t^{2\alpha - 1} K_t, \tag{15.123}$$

with similar equations for y and z (with β and γ replacing α, of course). For $\epsilon \neq t$, $\delta = t$, we find

$$K_{x,t} + K_{t,x} = 2\alpha t^{-1} K_x, \tag{15.124}$$

again with similar equations for y and z. Considering $\epsilon \neq t$, $\delta \neq t$, $\epsilon \neq \delta$, we have

$$K_{x,y} + K_{y,x} = 0, \tag{15.125}$$

again with the other two cyclic permutations. Finally, we let $\epsilon = \delta = t$ to find

$$K_{t,t} = 0. \tag{15.126}$$

We now set out to solve this set of ten coupled differential equations. We integrate equation (15.123) to find

$$K_x = \alpha t^{2\alpha-1} \left[\int K_t(x, y, z) dx + f(y, z, t) \right], \tag{15.127}$$

where f is a constant of integration. Differentiating equation (15.127) with respect to t yields

$$K_{x,t} = \frac{2\alpha - 1}{t} K_x + \alpha t^{2\alpha-1} f_{,t}, \tag{15.128}$$

which we use in equation (15.124):

$$\frac{2\alpha - 1}{t} K_x + \alpha t^{2\alpha-1} f_{,t} + K_{t,x} = \frac{2\alpha}{t} K_x. \tag{15.129}$$

We rearrange this and differentiate with respect to x, (noting that $f_{,tx} = f_{,xt} = 0$) and then use equation (15.123) for $K_{x,x}$, arriving at

$$K_{t,xx} = \alpha t^{2\alpha-2} K_t. \tag{15.130}$$

Equation (15.126) tells us that K_t is independent of time, so $K_{t,xx}$ must also be independent of time. Thus equation (15.130) implies $K_t = 0$ unless $\alpha = 0$ or $\alpha = 1$. By extending this argument to y and z, and using the conditions (15.119) and (15.120), we see that we must have $K_t = 0$ unless α, β, and γ are all equal to zero (which is just the trivial case of Minkowski space) or one of the three is equal to 1 while the other two vanish. We will treat the nontrivial special case separately, after considering all the remaining, more general, cases where $K_t = 0$.

$K_t = 0$

With $K_t = 0$, we have from equation (15.123) that $K_{x,x} = 0$ (and similarly for $K_{y,y}$ and $K_{z,z}$). Since $K_t = 0$, $K_{t,x} = 0$. Thus equation (15.124) simplifies to

$$K_{x,t}(y, z, t) = 2\alpha t^{-1} K_x(y, z, t), \qquad (15.131)$$

which has the solution

$$K_x = t^{2\alpha} \eta(y, z), \qquad (15.132)$$

where $\eta(y, z)$ is a function containing the y and z dependence of K_x. Similarly we find

$$K_y = t^{2\beta} \mu(x, z), \quad K_z = t^{2\gamma} \rho(x, y). \qquad (15.133)$$

Finally, we use the above in equation (15.125) to find the conditions

$$
\begin{aligned}
t^{2\alpha} \eta_{,y}(y, z) + t^{2\beta} \mu_{,x}(x, z) &= 0, & (15.134) \\
t^{2\alpha} \eta_{,z}(y, z) + t^{2\gamma} \rho_{,x}(x, y) &= 0, & (15.135) \\
t^{2\beta} \mu_{,z}(x, z) + t^{2\gamma} \rho_{,y}(x, y) &= 0. & (15.136)
\end{aligned}
$$

We must consider two cases:

1. $\alpha = \beta \neq \gamma$ (That is, choose any two of α, β, γ to be equal to each other but unequal to the third.)

   Then, in order to satisfy equations (15.135) and (15.136) at all times t, each term separately must equal zero. Thus, we must have that $\rho(x, y)$ is constant, $\eta(y, z) = \eta(y)$, and $\mu(x, z) = \mu(x)$. From equation (15.134), $\eta_{,y}(y) = -\mu_{,x}(x)$. This gives us the solutions

   $$\eta = hy + l, \quad \mu = -hx + n, \quad \rho = p, \qquad (15.137)$$

   where h, l, n, and p are constants.

2. $\alpha \neq \beta, \beta \neq \gamma, \alpha \neq \gamma$

Again, in order to satisfy equations (15.134), (15.135), and (15.136) at all times t, each function must be constant:

$$\eta = l, \quad \mu = n, \quad \rho = p, \qquad (15.138)$$

with l, n, and p constants.

Now that we have solved for η, ρ, and μ for all non-special cases, we may write a general solution for the Killing vectors. Consider the Killing *one-form*: $K = (0, \eta t^{2\alpha}, \mu t^{2\beta}, \rho t^{2\gamma})$, from (15.132) and (15.133). Evidently the Killing *vector* has the form

$$\begin{pmatrix} 0 \\ \eta \\ \mu \\ \rho \end{pmatrix}. \qquad (15.139)$$

We wish to find a set of basis vectors for each of the two cases. These sets follow directly from the η, μ, and ρ found in each case.

1. $\alpha = \beta \neq \gamma$

In this case, we have the three spatial translations, (15.122), and the rotation in the x-y plane:

$$\begin{pmatrix} 0 \\ y \\ -x \\ 0 \end{pmatrix}. \qquad (15.140)$$

2. $\alpha \neq \beta, \beta \neq \gamma, \alpha \neq \gamma$

In this case we have only the three spatial translations, (15.122).

$K_t \neq 0$

Finally, let us consider the case where $\alpha = 1$, $\beta = \gamma = 0$. By a process similar to that above, we again solve the ten coupled differential equations given by equations (15.123) through (15.126). This is a tedious process. The result is that we find ten Killing vectors (therefore space is flat!). Four of these are the three translations in x, y, and z and the rotation between y and z. The remaining six are given below:

$$\begin{pmatrix} -\sinh x \\ t^{-1}\cosh x \\ 0 \\ 0 \end{pmatrix}, \begin{pmatrix} -\cosh x \\ t^{-1}\sinh x \\ 0 \\ 0 \end{pmatrix}, \begin{pmatrix} -y\sinh x \\ t^{-1}y\cosh x \\ -t\sinh x \\ 0 \end{pmatrix}, \qquad (15.141)$$

and

$$\begin{pmatrix} -y\cosh x \\ t^{-1}y\sinh x \\ -t\cosh x \\ 0 \end{pmatrix}, \begin{pmatrix} -z\sinh x \\ t^{-1}z\cosh x \\ 0 \\ -t\sinh x \end{pmatrix}, \begin{pmatrix} -z\cosh x \\ t^{-1}z\sinh x \\ 0 \\ -t\cosh x \end{pmatrix}. \qquad (15.142)$$

These six were not immediately apparent from the metric. However, we could have made the substitution $V = t\cosh x$ and $U = t\sinh x$ into the metric, resulting in

$$ds^2 = -dV^2 + dU^2 + dy^2 + dz^2. \qquad (15.143)$$

In this metric, we have the ten Minkowski Killing vectors, which we may transform into our coordinates, duplicating the ten vectors we found above. Notice that we have now answered part (d): this space is indeed flat.

c) We assume that at least one of α, β, and γ is nonzero. The two conditions (15.119) and (15.120) then tell us that $\alpha + \beta + \gamma = 1$ and $\alpha^2 + \beta^2 + \gamma^2 = 1$. These two requirements together mean that the allowed set of α, β, γ corresponds to the intersection of the plane given by $\alpha+\beta+\gamma = 1$ with the unit sphere. Thus, we either have one variable equal to 1, with the other two equal to 0, or one of the variables must be negative. A negative value for α, β, or γ means that our spacetime is shrinking along one direction. If this were so, Hubble's law would not be observed in all directions. Also, the cosmic microwave background would appear *blue*-shifted in some spots, rather than uniformly red-shifted, and this is not observed. If one variable is equal to 1 while the others vanish, the microwave background would not be red-shifted at all in some directions, which also does not jibe with observations.

Chapter 16

Nuclear Physics—Solutions

Solution 7.1. a) The ground state of $^{205}_{81}$Tl has proton and neutron configuration

$$
\begin{aligned}
&\text{p } [82]\,(3s_{1/2})^{-1} \\
&\text{n } [126]\,(3p_{1/2})^{-2}\,,
\end{aligned}
\tag{16.1}
$$

where $[k]$ denotes a closed, inert core of k nucleons, and $(\cdots)^{-1}$ denotes a "hole" in that core. There are two neutron holes, which will pair so that their combined wavefunction has total spin zero and positive parity. Thus the properties of the ground state are determined by the lone proton hole, which has orbital angular momentum $l = 0$, parity $\pi = (-1)^l = +1$, and spin 1/2, so that $J^\pi = 1/2^+$.

The simplest possibility for the excited state is that the hole is excited from the $3s_{1/2}$ to the $2d_{3/2}$ shell (or equivalently, a proton is excited from the $2d_{3/2}$ shell to the $3s_{1/2}$). This shell has $l = 2$, so the parity is again positive. Therefore $J^\pi = 3/2^+$.

For $^{205}_{82}$Pb, the proton shell is closed, and we will assume it is inert. The neutron structure is

$$
\text{n } [126]\,(2f_{5/2})^{-1}\,(3p_{1/2})^{-2}\,.
\tag{16.2}
$$

Again, the $3p_{1/2}$ holes pair to spin zero, and the properties of the ground state arise from the lone neutron hole in the $2f_{5/2}$ shell. Now $l = 3$, so the parity is negative. Therefore $J^\pi = 5/2^-$.

There are many possibilities for excited states. One is to excite a $3p_{1/2}$ hole to the $2f_{5/2}$ shell, giving

$$n\,[126]\,(2f_{5/2})^{-2}\,(3p_{1/2})^{-1}\,. \tag{16.3}$$

Now it is the $2f_{5/2}$ holes that will pair, and we get $J^{\pi} = 1/2^-$. Alternatively, we can excite the $2f_{5/2}$ hole to the $3p_{3/2}$ shell, giving

$$n\,[126]\,(3p_{3/2})^{-1}\,(3p_{1/2})^{-2}\,, \tag{16.4}$$

which has $J^{\pi} = 3/2^-$.

b) $^{206}_{81}$Tl has the structure

$$\begin{aligned}
&p\,[82]\,(3s_{1/2})^{-1} \\
&n\,[126]\,(3p_{1/2})^{-1}\,.
\end{aligned} \tag{16.5}$$

The proton has $l = 0$ and the neutron has $l = 1$, so $\pi = -1$. The spin is hard to predict with confidence, as odd-odd systems are notoriously fickle, but it is certainly plausible that the two spin $1/2$'s could couple to total spin zero.

For $^{206}_{82}$Pb, again we have a filled, inert core of protons, and a neutron structure of

$$n\,[126]\,(3p_{1/2})^{-2}\,. \tag{16.6}$$

The holes will pair to give $J^{\pi} = 0^+$.

If we imagine exciting one hole to the $2f_{5/2}$ shell, we have to couple $j = 5/2$ to $j = 1/2$, and the resulting total spin will be either $J = 2$ or $J = 3$, both with positive parity. If we excite both $3p_{1/2}$ holes to the $2f_{5/2}$ shell, the spins will pair to give $J = 0$. Thus all observed states can be plausibly described within the shell model.

c) If there is unit density of neutrinos, the capture rate for the reaction $^{205}_{81}$Tl $+\nu \rightarrow$ $^{205}_{82}$Pb $+e^-$ is given by Fermi's golden rule:

$$\Gamma = \frac{2\pi}{\hbar}\,|H_{if}|^2\,\rho(E)\,, \tag{16.7}$$

where $\rho(E)$ is the phase space available to the final state particles with total energy E, and H_{if} is the matrix element of the interaction hamiltonian between the initial and final states. For a neutrino flux j_ν, the capture rate Γ_{cap} per ^{205}Tl atom is

$$\Gamma_{cap} = j_\nu \sigma = j_\nu \frac{\Gamma}{c} = j_\nu \frac{2\pi}{\hbar c} |H_{if}|^2 \rho(E), \tag{16.8}$$

with Γ as in equation (16.7) and σ the cross-section for neutrino capture. (Because the neutrino velocity is c, $\sigma = \Gamma/c$.)

It is hard to calculate H_{if}, as it is the matrix element of some complicated operator taken between the initial state of a single neutrino and a nucleus of Tl, and a final state of Pb and an electron. We will denote this by $|H_{if}|^2 = |\mathcal{M}|^2_{205}$. Now we have to use the rest of the data we are given.

Consider the decay ^{206}Tl $\rightarrow$ ^{206}Pb $+ e^- + \nu$. This is a standard beta-decay reaction, and the halflife is given by the well-known formula for the comparative halflife (for a derivation see, e.g., Cottingham and Greenwood):

$$ft_{1/2} = \frac{2\pi^3 \hbar^7 \ln 2}{|\mathcal{M}|^2_{206} m_e^5 c^4}. \tag{16.9}$$

Since we are given $\log ft_{1/2} = 5.2$, we can use the formula above to find $|\mathcal{M}|^2_{206}$, which we will then use to calculate $|\mathcal{M}|^2_{205}$.

The simplest and most important decay to consider is one in which a single proton or neutron decays, emitting a beta particle and a neutrino, and in the course of the decay is perhaps excited into a different nuclear orbital. In other words, the interaction is mediated by a *single particle operator*. However, if we compare the ground state configuration for ^{205}Tl, as given in part (a),

$$\begin{array}{l} \text{p } [82] \ (3s_{1/2})^{-1} \\ \text{n } [126] \ (3p_{1/2})^{-2} \ , \end{array} \tag{16.10}$$

with that of the $1/2^-$ state in ^{205}Pb,

$$\begin{array}{l} \text{p } [82] \\ \text{n } [126] \ (2f_{5/2})^{-2} \ (3p_{1/2})^{-1} \ , \end{array} \tag{16.11}$$

we see that in addition to a $2f_{5/2}$ neutron capturing a neutrino and being excited to a $3s_{1/2}$ proton orbital, a *second* $2f_{5/2}$ neutron needs to be excited up to the $3p_{1/2}$ shell. Because this is a two-particle process, the matrix element for this reaction will be very small, and hence the reaction rate will be negligibly small.

Fortunately, the question tells us to assume that the ground state of ^{205}Tl has a 10% admixture of the configuration

$$\begin{aligned} &\text{p } [82] \, (3s_{1/2})^{-1} \\ &\text{n} \, [126] \, (2f_{5/2})^{-2} \ . \end{aligned} \tag{16.12}$$

From this state a single neutron in the $3p_{1/2}$ shell can capture a neutrino and end up as a proton in the $3s_{1/2}$ shell. Thus the matrix element for this capture will dominate the reaction rate.

We can now compare the capture, ^{205}Tl $\rightarrow$ ^{205}Pb* (where the asterisk refers to the excited state),

$$\begin{aligned} &\text{p } [82] \, (3s_{1/2})^{-1} \ \rightarrow \ [82] \\ &\text{n} \, [126] \, (2f_{5/2})^{-2} \ \rightarrow \ [126] \, (2f_{5/2})^{-2} \, (3p_{1/2})^{-1} \ , \end{aligned} \tag{16.13}$$

with the decay ^{206}Tl $\rightarrow$ ^{206}Pb,

$$\begin{aligned} &\text{p } [82] \, (3s_{1/2})^{-1} \ \rightarrow \ [82] \\ &\text{n} \, [126] \, (3p_{1/2})^{-1} \ \rightarrow \ [126] \, (3p_{1/2})^{-2} \ . \end{aligned} \tag{16.14}$$

In both cases, a $3p_{1/2}$ neutron decays to a $3s_{1/2}$ proton, and all other nucleons are undisturbed. Thus we would expect the matrix elements for these reactions to be approximately equal. Since the ^{205}Tl ground state is only in the configuration (16.12) 10% of the time, the matrix element for the ^{205}Tl ground state is only 10% of that of ^{206}Tl,

$$|\mathcal{M}|^2_{205} \approx 0.1 \times |\mathcal{M}|^2_{206} \ . \tag{16.15}$$

We now need to evaluate the flux of neutrinos from the sun. We are told that the energy flux is 0.14 W cm^{-2} = 8.8×10^{11}MeV cm^{-2}s^{-1}, and that there is one neutrino for every 13.1 MeV. Thus the neutrino flux j_ν is

$$\begin{aligned} j_\nu &= \frac{(8.8 \times 10^{11}\text{MeV cm}^{-2})}{(13.4 \text{ MeV})} = 6.7 \times 10^{10}\text{cm}^{-2}\text{s}^{-1} \\ &= 6.7 \times 10^{-16}\text{fm}^{-2}\text{s}^{-1}. \end{aligned} \tag{16.16}$$

The last quantity we need is the density-of-states or phase space factor. Since the Pb nucleus is far more massive than the electron, we can make the very good approximation that the nucleus will "take care" of momentum conservation while remaining essentially at rest — i.e., it will carry a negligible fraction of the kinetic energy. We also approximate the outgoing electrons by plane waves (of course this is not completely correct because of the attraction between the electron and the nucleus): then there is one state per volume $(2\pi\hbar)^3$ in momentum space.

Given these simplifications, the density of states is given by the expression

$$\rho(E)\,dE = \frac{d^3p}{(2\pi\hbar)^3} = \frac{4\pi p^2\,dp}{(2\pi\hbar)^3} = \frac{E\sqrt{E^2 - m^2c^4}\,dE}{2\pi^2\hbar^3c^3}\,, \tag{16.17}$$

where now E is the energy of the electron, and we have used the relativistic formula $E^2 = m^2c^4 + p^2c^2$ to eliminate momentum in favor of energy. Assuming a neutrino energy of 0.26 MeV, the total energy of the electron is the sum of the neutrino energy and the electron rest mass, minus the sum of Q_{EC} and the excitation energy of $^{205}\mathrm{Pb}^*$:

$$E = (0.26 - 0.06 - 0.002 + 0.511)\ \mathrm{MeV} \approx 0.71\ \mathrm{MeV}\,. \tag{16.18}$$

If we use this to evaluate the density of states we find

$$\rho(E) \approx \frac{(1.8 \times 10^{-2}\ \mathrm{MeV}^2)}{(\hbar c)^3}\,. \tag{16.19}$$

Combining equations (16.8), (16.9), (16.15), (16.16), and (16.19), we can evaluate the capture rate Γ_{cap} per $^{205}\mathrm{Tl}$ atom:

$$\Gamma_{cap} = j_\nu \times \frac{2\pi}{\hbar c} \times |\mathcal{M}|^2_{205} \times \frac{(1.8 \times 10^{-2}\mathrm{MeV}^{-2})}{(\hbar c)^3}$$

$$= 1.5 \times 10^{-36}\mathrm{s}^{-1}\,. \tag{16.20}$$

The rate for a single atom of $^{205}\mathrm{Pb}$ to decay to $^{205}\mathrm{Tl}$ is found using the halflife given in the question:

$$\Gamma_{decay} = \frac{1}{\tau} = \frac{\ln 2}{(1.4 \times 10^7) \times (3 \times 10^7\ \mathrm{s})} \approx 1.7 \times 10^{-15}\ \mathrm{s}^{-1}\,. \tag{16.21}$$

If we have n_{eq} atoms of ^{205}Pb and n_{Tl} atoms of ^{205}Tl, the number decaying per second is $n_{eq} \times \Gamma_{decay}$, and in equilibrium this is equal to the production rate:

$$n_{eq} \times \Gamma_{decay} = n_{Tl} \times \Gamma_{cap} . \tag{16.22}$$

The number of atoms in one gram of thallium is Avogadro's number divided by the atomic weight, $n_{Tl} = (6 \times 10^{23}/205)$ atoms. Therefore the equilibrium number of atoms of ^{205}Pb is $n_{eq} \approx 2.5$ atoms per gram of ^{205}Tl.

Solution 7.2.   a) We assume that we can approximate the nucleus by an inert core of $A - 1$ nucleons, arranged in pairs of total spin zero, orbited by a single nucleon of spin $1/2$. The orbital angular momentum is l, and total angular momentum is $I = l \pm \frac{1}{2}$. The magnetic moment operator of the nucleus is the sum of the orbital and intrinsic magnetic moments of the nucleon,

$$\boldsymbol{\mu}_I = \boldsymbol{\mu}_l + \boldsymbol{\mu}_s = \frac{\mu_N}{\hbar}\left(g_l \mathbf{l} + g_s \mathbf{s}\right) . \tag{16.23}$$

In these equations, $\mu_N = e\hbar/2m_p$ is the nuclear magneton and $g_l = 1$ for the proton and 0 for the neutron. Similarly, $g_s \approx 5.6$ and -3.8 respectively.

The expectation value of $\boldsymbol{\mu}_I$ must be parallel to the total spin $\mathbf{I} = \mathbf{l} + \mathbf{s}$ since this is the only possible direction: $\langle \boldsymbol{\mu}_I \rangle = k \langle \mathbf{I} \rangle$. We can use the Wigner-Eckart theorem to eliminate k and find the relationship

$$\langle \boldsymbol{\mu}_I \rangle = \frac{\langle \boldsymbol{\mu}_I \cdot \mathbf{I} \rangle}{I(I+1)\hbar^2} \langle \mathbf{I} \rangle . \tag{16.24}$$

The magnetic moment is defined as the expectation value of the z-component of $\boldsymbol{\mu}_I$ in a state with $I_z = I\hbar$,

$$\mu_I \equiv \frac{\langle \boldsymbol{\mu}_I \cdot \mathbf{I} \rangle}{(I+1)\hbar} . \tag{16.25}$$

We can evaluate this by rewriting equation (16.23) in the form

$$\mu_I = \frac{\mu_N}{\hbar}\left(\frac{1}{2}(g_l + g_s)(\mathbf{l} + \mathbf{s}) + \frac{1}{2}(g_l - g_s)(\mathbf{l} - \mathbf{s})\right), \qquad (16.26)$$

and taking the scalar product of this with $\mathbf{I}$ to get

$$\mu_I \cdot \mathbf{I} = \frac{\mu_N}{\hbar}\left(\frac{1}{2}(g_l + g_s)\mathbf{I}^2 + \frac{1}{2}(g_l - g_s)(\mathbf{l}^2 - \mathbf{s}^2)\right). \qquad (16.27)$$

By taking the expectation value of this and simplifying, we find an expression for the gyromagnetic ratio $g_I = \mu_I/(\mu_N I)$:

$$g_I = \frac{1}{2}(g_l + g_s) + \frac{1}{2}(g_l - g_s)\frac{l(l+1) - 3/4}{I(I+1)}. \qquad (16.28)$$

For the case $I = l - 1/2$ this simplifies to

$$g_I = g_l + \frac{g_l - g_s}{2(I+1)}, \qquad (16.29)$$

and for $I = l + 1/2$,

$$g_I = g_l - \frac{g_l - g_s}{2I}. \qquad (16.30)$$

b) There are many ways (find your own) of remembering the sequence of energy levels:

$$1s_{1/2} \quad | \quad 1p_{3/2}\ 1p_{1/2} \quad | \quad 1d_{5/2}\ 2s_{1/2}\ 1d_{3/2} \quad | \quad 1f_{7/2}. \qquad (16.31)$$
$$\phantom{1s_{1/2}} \quad 2 \qquad\qquad 8 \qquad\qquad\qquad\qquad 20$$

In this table, vertical lines denote "magically" closed shells, for which the excitation energy of a nucleon to the next shell is unusually high.

c) The $^{17}_{8}O_9$ nucleus has an unpaired neutron in the $1d_{5/2}$ shell. The neutron has $l = 2$ and $I = 5/2 = l + 1/2$, so inserting the relevant values of $g_l = 0$ and $g_s = -3.8$ into equation (16.30), we find the value $g_I = -0.76$. This corresponds to a magnetic moment of

$$\mu_I = \frac{5}{2}g_I\mu_N = -1.9\mu_N. \qquad (16.32)$$

For the case of $^{23}_{11}\text{Na}_{12}$, there is an unpaired proton in the $1d_{5/2}$ shell, which has $g_l = 1$ and $g_s = 5.6$. Therefore $g_I = 1.9$ and the magnetic moment is $\mu_I = 4.8\mu_N$. Similarly, $^{45}_{21}\text{Sc}_{24}$ has an unpaired proton in the $1f_{7/2}$ shell, so $l = 3$, $I = l + 1/2 = 7/2$, $g_I = 1.7$, and $\mu_I = 5.8\mu_N$.

d) To calculate the electric quadrupole moment of $^{15}_{7}\text{N}_8$, we need its proton configuration:

$$\left[(1s_{1/2})^2 (1p_{3/2})^4 (1p_{1/2})^2\right] (1p_{1/2})^{-1}, \tag{16.33}$$

where $[\cdots]$ denotes a magically closed core, and $(\cdots)^{-1}$ means a hole of charge $-e$. In the extreme single particle model, we assume that the core does not contribute to the quadrupole moment, which is defined as

$$Q = \frac{1}{e}\sum\langle q(3z^2 - r^2)\rangle. \tag{16.34}$$

Here the sum is over all particles (protons or holes) contributing to the moment, q is the particle charge, and the expectation value is taken in the state with $I_z = I$. In the case of interest, namely that of a single particle orbiting a sperically symmetric core, the wavefunction will separate into a product of angular and radial components. We can therefore rewrite Q as

$$Q = \frac{1}{e}\sum\langle qr^2\rangle\langle(3\cos^2\theta - 1)\rangle, \tag{16.35}$$

where θ is the polar angle. It is easy to evaluate the angular part of the expectation value. In spherical coordinates, $\cos^2\theta = z^2/r^2$. In an analogous fashion, the expectation value of $\cos^2\theta$ in a state of angular momentum $|I\,I_z\rangle$ can be written

$$\langle I\,I_z|\cos^2\theta|I\,I_z\rangle = \left\langle\frac{I_z^2}{\mathbf{I}^2}\right\rangle = \frac{\langle I_z^2\rangle}{I(I+1)}. \tag{16.36}$$

For $I_z = I$, we find

$$\langle 3\cos^2\theta - 1\rangle = \frac{3I}{I+1} - 1 = \frac{2I-1}{I+1}. \tag{16.37}$$

The radial expectation value can be approximated in terms of typical nuclear radii, which are given by the empirical formula $r \approx 1.1A^{1/3}$ fm.

For the case of $^{15}_{7}N_8$, the total spin of the nucleus is $I = 1/2$, and we see that $Q = 0$. For $^{11}_{5}B_6$ the proton configuration is

$$\left[(1s_{1/2})^2 \, (1p_{3/2})^4 \right] (1p_{3/2})^{-1}, \tag{16.38}$$

so we have a hole in a state of $I = 3/2$. Therefore the quadrupole moment is

$$Q = -\langle r^2 \rangle \frac{3-1}{3/2+1} = -\frac{4}{5} \langle r^2 \rangle . \tag{16.39}$$

Inserting a value of $r \approx 3 \times 10^{-13}$ cm, we find a quadrupole moment of

$$Q \approx -7 \times 10^{-26} \text{ cm}^2 . \tag{16.40}$$

Solution 7.3. a) Since $^{14}_{8}O_6$ is an even-even nucleus, $J^\pi = 0^+$. The nucleus has an excess of two protons so that the z-component of the isospin T_z is one. The lowest energy state will have the lowest possible isospin, which means that the isospin is $T = 1$.

Superallowed Fermi transitions occur for $\Delta J = \Delta \pi = \Delta T = 0$. Therefore the 2.3 MeV state of ^{14}N also has $J^\pi = 0^+$ and $T = 1$.

b) In the shell model, the proton configuration for ^{14}O is $(1s_{1/2})^2(1p_{3/2})^4(1p_{1/2})^2$. Since the β-decay is a superallowed Fermi decay, the squared matrix element for the decay of a single proton is just G_F^2, where G_F is the Fermi constant. Either one of the protons in the $1p_{1/2}$ shell can β-decay to a neutron to form ^{14}N, so the square of the matrix element is

$$|\langle M \rangle|^2 = 2G_F^2 . \tag{16.41}$$

c) The proton and neutron configurations for ^{14}N are both $(1p_{1/2})^1$. The parity of the state is given by the product of the parities of the proton and neutron wavefunctions. Therefore the parity must be positive. The spin J could be zero or one. However since the 2.3 MeV state has $J = 0$, the impossibility of a $J = 0 \rightarrow J = 0$ γ-transition implies that $J = 1$ for the ground state. Finally, since there are equal numbers of protons and neutrons, $T_z = 0$, so the ground state will have $T = 0$. The β decay from the ground state of ^{14}O to the ground state of ^{14}N has $\Delta T = \Delta J = 1$, $\Delta \pi = 0$. Therefore the decay is an allowed Gamow-Teller transition. Since ^{14}C and ^{14}O are mirror nuclei, not only is the β decay from ^{14}C to ^{14}N also an allowed Gamow-Teller transition, but the two β decays have nearly the same matrix element.

d) The phase-space factor for β decay is proportional to E_0^5, where E_0 is the end-point energy, provided $E_0 \gg m_e$. The end-point energy is simply the total available kinetic energy from the decay, or

$$E_0 = M(A, Z) - M(A, Z - 1) - m_e , \qquad (16.42)$$

where $M(A, Z)$ is the mass of a nucleus of atomic number A and proton number Z. The semi-empirical mass formula is

$$M(A, Z) = Z m_p + N m_n - aA + bA^{2/3} + c\frac{(N - Z)^2}{A} + d\frac{Z^2}{A^{1/3}} + \frac{\delta}{A^{3/4}}.$$
$$(16.43)$$

Approximate values for the constants appearing in this expression are (in MeV): $a = 16$, $b = 18$, $c = 24$, $d = 0.7$, and $\delta = -33$ for even-even nuclei, $\delta = 0$ for odd-even nuclei, and $\delta = +33$ for odd-odd nuclei (see Burcham, Chapter 6). If we assume that $Z = A/2$, then for reasonably large A, it is easily verified using the constants given above that the Coulomb term (i.e., the term with the coefficient d) dominates the mass difference between the nuclei. The difference in this term for a nucleus with Z protons and for one with $(Z - 1)$ protons is about $2dZ/A^{1/3}$, so that if Z is proportional to A then E_0 is proportional to $A^{2/3}$. Since the rate of decay is proportional to E_0^5, the halflife $t_{1/2}$ is proportional to $A^{-10/3}$.

e) The question gives the halflife for ^{14}O as 70 s. Using the results from part (d), if $A = 54$ we would expect that $t_{1/2} = 70(14/54)^{10/3} = 0.78$ s. Our estimate is off by a factor of about four from the real decay time.

Solution 7.4. A. a) We wish to find the number of neutrinos that are produced in the sun for a given luminosity, and then use the sun's luminosity to calculate a neutrino flux. The solar pp cycle is:

$$p + p \;\; \rightarrow \;\; d + e^+ + \nu_e + Q_1, \qquad\qquad (16.44)$$

$$d + p \;\; \rightarrow \;\; {}^3_2\text{He} + \gamma + Q_2, \qquad\qquad (16.45)$$

$$ {}^3_2\text{He} + {}^3_2\text{He} \;\; \rightarrow \;\; {}^4_2\text{He} + 2p + Q_3, \qquad\qquad (16.46)$$

where d is a deuteron. For our purposes this cycle is equivalent to

$$4p \rightarrow {}^4_2\text{He} + 2e^+ + 2\nu_e + Q_{total}. \qquad\qquad (16.47)$$

Using the atomic mass excesses we see that

$$
\begin{aligned}
Q_{total} \;\; &= \;\; 4[m({}^1_1\text{H}) - m(e^-)] - [m({}^4_2\text{He}) - 2m(e^-)] - 2m(e^+) \\
&\approx \;\; 24.7 \text{ MeV}. \qquad\qquad\qquad\qquad\qquad\qquad\qquad\qquad (16.48)
\end{aligned}
$$

The positrons created will each annihilate with electrons, releasing an additional 2 MeV of energy. Thus, one neutrino is produced for approximately each 13 MeV of luminosity.

The solar luminosity at the earth's orbit is given as 1.4 kW/m², so we have a neutino flux f at the earth's orbit given by

$$
\begin{aligned}
f \;\; &\approx \;\; \left(1.4 \text{ kW/m}^2\right) \left(\frac{1}{13 \times 10^6 \text{ eV}}\right) \left(\frac{1 \text{ eV}}{1.6 \times 10^{-19} \text{ J}}\right) \\
&\approx \;\; 7 \times 10^{14} \text{ m}^{-2}\text{s}^{-1}. \qquad\qquad\qquad\qquad\qquad\qquad (16.49)
\end{aligned}
$$

b) There is a spectrum of neutrino energies because the neutrino shares the energy of the decay with an electron. From Fermi's golden rule, the rate for a decay to a neutrino with energy between E_ν and $E_\nu + dE_\nu$ is

$$d\Gamma = \frac{2\pi}{\hbar} |M| \left(\frac{d\rho}{dE_\nu}\right) dE_\nu, \qquad\qquad (16.50)$$

where M is the matrix element governing the decay (16.44) and ρ is the final state's phase-space factor. We assume that the matrix element is independent of the energy of the outgoing neutrino. Implicit in the matrix element is conservation of momentum and energy. Since the deuteron is much more massive than the other two decay products, it will carry very little of the kinetic energy while effectively taking care of momentum conservation. Therefore the E_ν dependence of (16.50) reduces to

$$d\Gamma \propto n_\nu(E_\nu)n_e(E_0 - E_\nu)dE_\nu, \qquad (16.51)$$

where E_0 is equal to Q_1, the energy of the decay, and n_ν and n_e are the densities of states of the neutrino and electron, respectively. We find these densities of states by writing the total number of states for either particle in a volume of k-space,

$$N = \frac{V}{(2\pi)^3}\frac{4}{3}\pi k^3, \qquad (16.52)$$

and differentiating with respect to k, giving

$$dN = \frac{V}{(2\pi)^3}4\pi k^2 dk. \qquad (16.53)$$

To find $n_\nu(E) \equiv dN/dE$ we use the dispersion relation for a massless neutrino, $E = \hbar k c$, to find $dE = \hbar c \, dk$, which gives

$$n_\nu(E_\nu) = \frac{V}{(2\pi\hbar c)^3}4\pi E_\nu^2. \qquad (16.54)$$

For the electron, we use the dispersion relation $E_e^2 = m_e^2 c^4 + \hbar^2 k^2 c^2$, and again we calculate the differential dk in terms of dE and substitute into the expression for $n(k)$ to find

$$n_e(E_e) = \frac{V}{(2\pi\hbar c)^3}4\pi E_e(E_e^2 - m_e^2 c^4)^{\frac{1}{2}}. \qquad (16.55)$$

Substituting $E_e = E_0 - E_\nu$ and using n_e and n_ν in (16.51) gives us the shape of the spectrum:

$$\frac{d\Gamma}{dE_\nu} \propto E_\nu^2(E_0 - E_\nu)[(E_0 - E_\nu)^2 - m_e^2 c^4]^{\frac{1}{2}}. \qquad (16.56)$$

c) The energy available to the neutrino is found by using the atomic mass excesses and the first reaction of the pp cycle, (16.44). This gives $Q_1 = 0.42$ MeV. If we estimate that the neutrino gets about half of this, the percentage of the total energy of the cycle carried off by the neutrino is

$$\frac{0.2}{13} \approx 1.5\%. \qquad (16.57)$$

B. a) In the radioactive decay of heavy elements, the α decays make the nuclei neutron-rich, so the nuclei convert neutrons to protons via the reaction

$$n \to p + e^- + \bar{\nu}_e, \qquad (16.58)$$

and antineutrinos are produced.

b) From the given values of the temperature gradient of the earth's crust and the thermal conductivity of granite λ, we can find the heat flux given by $J_Q = \lambda dT/dz$. It follows that

$$J_Q = 2 \times 10^{15} \text{ MeV m}^{-2} \text{ hr}^{-1}. \qquad (16.59)$$

Now consider the source of energy suggested, the chain of decays from $^{232}_{90}$Th to $^{208}_{82}$Pb. We need siz α's and four β's to make this chain, which means four $\bar{\nu}_e$'s are released. Using the atomic mass excesses, we find that about 10 MeV of energy is produced for each antineutrino. Putting this together with J_Q, (16.59), gives a flux of antineutrinos of

$$f = \frac{2 \times 10^{15} \text{ MeV m}^{-2} \text{ hr}^{-1}}{10 \text{ MeV}} = 2 \times 10^{14} \text{ m}^{-2} \text{ hr}^{-1}. \qquad (16.60)$$

C. a) When a heavy nucleus fissions, the daughter nuclei are neutron-rich, and again antineutrinos are produced.

b) A one gigawatt reactor running with a 10% conversion efficiency produces a power of 10 GW $= 6 \times 10^{22}$ MeV/s through nuclear reactions. We will estimate that there are approximately ten antineutrinos per fission and about 200 MeV released per fission. (See Cottingham and Greenwood for a more detailed discussion.) With these numbers, 100 meters from the reactor there will be a flux f of antineutrinos of

$$f \approx \frac{6 \times 10^{22} \text{ MeV/s}}{20 \text{ MeV}} \frac{1}{4\pi(100 \text{ m})^2} \approx 2 \times 10^{16} \text{ m}^{-2} \text{ s}^{-1}. \qquad (16.61)$$

Solution 7.5.   a) The transition from the $|J^\pi = 3/2^-, T = 3/2\rangle$ state at 15 MeV directly to the ground state with $|J^\pi = 1/2^-, T = 1/2\rangle$ has $\Delta J = 1$ and no parity change, which means this is an $M1$ or $E2$ transition. To find the decay width for this transition in ^{13}N, we will compare it to the analogous decay in ^{13}C, since the nuclei are isobaric analogs. In the extreme single particle model, the ground state proton configuration of ^{13}N is

$$(1s_{1/2})^2(1p_{3/2})^4(1p_{1/2})^1, \qquad (16.62)$$

which is identical to the ground state neutron configuration of ^{13}C. The $|3/2^-, \ 3/2\rangle$ state of ^{13}N has a proton excited to the $1p_{1/2}$ level so that the proton configuration is

$$(1s_{1/2})^2(1p_{3/2})^3(1p_{1/2})^2, \qquad (16.63)$$

which is of course identical to the neutron configuration for the excited $|3/2^-, \ 3/2\rangle$ state in ^{13}C.

We see from these configurations that a proton is being promoted from a level with $j = l + 1/2$ to $j = l - 1/2$. The selection rules tell us that the $M1$ transition will dominate the decay, with a decay width

$$\Gamma_{\gamma 0} \propto |\langle \psi_f | \mu | \psi_i \rangle|^2 (\Delta E)^3, \qquad (16.64)$$

where ΔE is the energy of the transition, μ is the magnetic dipole operator, $|\psi_i\rangle$ is the initial configuration of protons and neutrons and $|\psi_f\rangle$ is the final-state configuration. In the extreme single particle model, the magnetic dipole operator is simply $\mu = \mu_N(g_s \mathbf{s} + g_l \mathbf{l})/\hbar$, where the angular momentum operator $\mathbf{l}$ and the spin operator $\mathbf{s}$ act only on the excited nucleon (i.e., the proton in the case of ^{13}N and the neutron in the case of ^{13}C). For a proton, $g_s \approx 5.6$ and $g_l = 1$, while for a neutron, $g_s \approx -3.8$ and $g_l = 0$. The matrix element appearing in (16.64) can now be partially evaluated to yield

$$
\begin{aligned}
|\langle\psi_f|\,\mu\,|\psi_i\rangle|^2 &= \left|\langle\psi_f|\frac{\mu_N}{2\hbar}[(g_s+g_l)\mathbf{j}+(g_s-g_l)(\mathbf{s}-\mathbf{l})]|\psi_i\rangle\right|^2 \\
&= \left|\frac{\mu_N}{2\hbar}(g_s-g_l)\langle\psi_f|(\mathbf{s}-\mathbf{l})|\psi_i\rangle\right|^2,
\end{aligned} \tag{16.65}
$$

where we have used $\langle\psi_f|\mathbf{j}|\psi_i\rangle = 0$. The matrix element $\langle\psi_f|\mathbf{s}-\mathbf{l}|\psi_i\rangle$ will be the same for the two decays; the only difference that arises is from the values of g_s and g_l. Therefore, the ratio of the squared matrix elements of ^{13}N and ^{13}C is $(5.6-1)^2/(3.8)^2 = 1.47$. Let us denote the partial widths of the transitions in nitrogen and carbon by $\Gamma_{\gamma_0}^N$ and $\Gamma_{\gamma_0}^C$, and the energy differences in each case as ΔE_N and ΔE_C. Then,

$$
\Gamma_{\gamma_0}^N \approx \Gamma_{\gamma_0}^C (1.47)\left(\frac{\Delta E_N}{\Delta E_C}\right)^3 \approx 39 \text{ eV}. \tag{16.66}
$$

b) The ground state neutron configuration of ^{13}C (also the proton configuration of ^{13}N) is given by (16.62), while the $1/2^+$ excited state has neutron configuration

$$
(1s_{1/2})^2(1p_{3/2})^4(2s_{1/2})^1. \tag{16.67}
$$

The explanation for the energy difference is fairly straightforward, with one subtlety. In going from a $1p_{1/2}$ to a $2s_{1/2}$ level, the charge distribution of the proton in nitrogen becomes "spread out," lowering the Coulomb energy of the nucleus. For carbon, a neutron is moved, leaving the electrostatic energy unchanged. This effect is always present in analog nuclei, but in this case there is a considerable enhancement due to the presence of the nearby unbound state of ^{13}C $+p$, which distorts the nuclear wavefunction and increases the average distance of the proton from the nucleus. (This is known as the Thomas-Ehrman effect.)

For completeness, we should point out that the reason there is no analogous energy difference between the 15 MeV excited states is that the corresponding unbound states would be a proton or neutron together with a ^{12}C core of isospin 1. These states are not shown, but in fact they lie several MeV above the 15 MeV bound states.

c) It is surprising that the two states have the same width because we would expect the decay rate to be proportional to some power of the excitation energy, and the two states have such different energies. Here, the transition is $E1$, and the rate will go as $(\Delta E)^3$. In addition, we need to take into account the matrix element of the dipole operator. If we assume that the spatial distributions of the nucleons are the same for the analog states in the two nuclei, and note that the effective charges of the valence nucleons, $(N/A)e$ for a proton and $-(Z/A)e$ for a neutron, have nearly the same magnitudes, then the matrix elements will be approximately equal. (See, for example, Hornyak, Chapter 7, for an explanation of "effective charge.") By arguments similar to those used in part (a), we expect

$$\frac{\Gamma_{\gamma_0}^N}{\Gamma_{\gamma_0}^C} \approx \left(\frac{\Delta E_N}{\Delta E_C}\right)^3 \approx 0.45 , \qquad (16.68)$$

rather than a ratio of unity.

However, in part (b) we already assumed that the wavefunction of ^{13}N is distorted, so that the proton is at a greater mean radius. This could have the effect of enhancing the dipole moment of the nucleus and thus increasing its decay rate.

B. a) In this part of the question, the Coulomb interaction is mixing the states of definite isospin.

Let us use $|+\rangle$ to designate the state at energy $E_+ = 5.67$ MeV and $|-\rangle$ for the state at $E_- = 5.60$ MeV. We will write these two orthonormal states in the following form:

$$\begin{aligned}
|+\rangle &= \cos\theta|T=0\rangle + \sin\theta|T=1\rangle, \\
|-\rangle &= \sin\theta|T=0\rangle - \cos\theta|T=1\rangle. \qquad (16.69)
\end{aligned}$$

Notice that for two-state mixing we may adjust the overall phases of the wavefunctions so that all the coefficients of the mixing matrix are real. The states $|+\rangle$ and $|-\rangle$ are the eigenstates of the Hamiltonian $\mathcal{H}$, by definition, so that we may write

$$\begin{aligned} \langle +|\mathcal{H}|+\rangle = E_+, \quad \langle -|\mathcal{H}|+\rangle = 0, \\ \langle +|\mathcal{H}|-\rangle = 0, \quad \langle -|\mathcal{H}|-\rangle = E_-. \end{aligned} \tag{16.70}$$

The mixing-matrix element $\mathcal{M}$ which we seek is

$$\mathcal{M} = \langle T = 0|\mathcal{H}|T = 1\rangle. \tag{16.71}$$

Upon inverting equations (16.69) we find

$$\begin{aligned} |T = 0\rangle = \cos\theta|+\rangle + \sin\theta|-\rangle, \\ |T = 1\rangle = \sin\theta|+\rangle - \cos\theta|-\rangle. \end{aligned} \tag{16.72}$$

It follows that the matrix element is

$$\mathcal{M} = (E_+ - E_-)\cos\theta\sin\theta = \frac{1}{2}(E_+ - E_-)\sin 2\theta. \tag{16.73}$$

Since isospin is conserved in alpha decay, only the $|T = 0\rangle$ component of $|+\rangle$ contributes to its partial width $\theta^2_{\alpha_+} = 0.09$ for decay into the isospin zero state $^{14}N + \alpha$, and similarly for the $|-\rangle$ state. Thus the partial widths are proportional to $\cos^2\theta$ and $\sin^2\theta$, respectively. There are other factors, due to phase space and differences in the wave functions, but we will ignore these. We find $\tan^2\theta = 0.09/0.17$, so that $\sin 2\theta \approx 0.95$, which we use in equation (16.73) to find

$$\mathcal{M} \approx 0.033 \text{ MeV.} \tag{16.74}$$

b) We note that each of the four transitions may proceed as $E1$, so that the energy dependence of each is $(\Delta E)^3$. Next, there is a selection rule for electric dipole transitions, requiring $\Delta T = \pm 1$ (see Hornyak), so only the $T = 0$ term can decay to the 1.04 MeV state. Therefore, the ratio of the partial width of the $|+\rangle$ state to that of the $|-\rangle$ state is

$$\left(\frac{\Gamma_\gamma^+}{\Gamma_\gamma^-}\right) \approx \cot^2\theta \left(\frac{\Delta E_+}{\Delta E_-}\right)^3 \approx 2.0, \tag{16.75}$$

for the decay to the 1.04 MeV state. Similarly, only the $T = 1$ component can decay to the ground state. Therefore, the ratio of the partial
widths for the decay to the ground state is

$$\left(\frac{\Gamma_\gamma^+}{\Gamma_\gamma^-}\right) \approx \tan^2\theta \left(\frac{\Delta E_+}{\Delta E_-}\right)^3 \approx 0.55\,. \tag{16.76}$$

Solution 7.6. a) The bulk-binding contribution to the nuclear binding
energy is proportional to the volume. The surface term, as the name
suggests, is proportional to the surface area of the nucleus, just as for
a drop of liquid the surface energy term will reduce the binding energy.
The Coulomb contribution is proportional to the square of the charge
on the nucleus and inversely proportional to its radius. In turn, the
volume, surface area, and radius of the nucleus are proportional to A,
$A^{2/3}$, and $A^{1/3}$, respectively. So the total binding energy E_B is

$$E_B \sim a_V A - a_S A^{2/3} - a_C Z^2/A^{1/3}\,. \tag{16.77}$$

b) To find a value of a_C we recall that the self-energy of a sphere of
charge Ze and radius $1.1A^{1/3}$ fm is

$$E = \frac{3}{5}\frac{Z^2 e^2}{1.1A^{1/3}\text{ fm}} = \left(\frac{3e^2}{5.5\text{ fm}}\right)\frac{Z^2}{A^{1/3}} \tag{16.78}$$

for large Z. Using $\hbar c \approx 197$ MeV-fm, we evaluate the term in paren-
thesis to find $a_C = 0.8$ MeV. The observed value is $a_C = 0.7$ MeV.

c) A rough justification of this term comes from the Pauli exclusion
principle, which suggests that the lowest energy nucleus for a given A
will be the one for which $N = Z$. So we want $N - Z = A - 2Z \approx 0$.
This term must come in quadratically, because its sign doesn't matter
(excess neutrons and excess protons contribute similar energies). By
considering a very asymmetric nucleus (with $Z \approx 0$), we can see that

the symmetry term should be a bulk term, proportional to the volume and hence to A, which requires

$$E \sim \frac{a_A(A - 2Z)^2}{A}.$$

(16.79)

From the condition of beta stability, we have (holding A fixed and combining the symmetry term with the expression for E_B given in (16.77))

$$\frac{\partial E_B}{\partial Z} = 0 = -\frac{2a_c Z}{A^{1/3}} - \frac{-4a_A(A - 2Z)}{A},$$

(16.80)

which, evaluated at $Z = 125$ and $A = 52$, gives $a_A/a_C \approx 30$.

d) The energy released in fission is $E_B(A, Z) - 2E_B(A/2, Z/2)$ for a symmetric decay. The contributions of the volume and the symmetry terms to this difference vanish identically. The surface term contributes an energy

$$\Delta_S = -a_S \left(A^{2/3} - 2 \left(\frac{A}{2} \right)^{2/3} \right) = 0.26 a_S A^{2/3} = 10.7 a_S.$$

(16.81)

The Coulomb contribution is

$$\Delta_C = -a_C \left(\frac{Z^2}{A^{1/3}} - 2 \left(\frac{Z}{2} \right)^2 \left(\frac{2}{A} \right)^{1/3} \right) = -0.37 a_C \frac{Z^2}{A^{1/3}} = -577 a_C.$$

(16.82)

Using the value for a_S given in the question, $a_S \approx 17$ MeV and $a_C \approx 0.8$ MeV from part (b), we find a total energy release of 280 MeV, which seems reasonable.

The dominance of the Coulomb force in this result shows that fission is predominantly a manifestation of electrostatic energy.

Solution 7.7.   a) Asking for the multipolarity of the radiation is the same as asking for its angular momentum quantum number. The rules for addition of angular momentum require that

$$|J_i - J_f| \leq L \leq |J_i + J_f|,$$

(16.83)

where J_i and J_f are the angular momenta of the initial and final states, and L is the multipolarity of the radiation field. Since we are considering a case where $J_i = 1$ and $J_f = 0$, it is evident that $L = 1$.

b) Consider the emission of a photon in a transition from an initial state with J_i to a final state with J_f, in an L-pole radiation field. The angular distribution of this field is given by

$$w(\theta) \propto \sum p(m_i)|\langle J_i m_i|J_f m_f LM\rangle|^2 |X_L^M|^2 , \qquad (16.84)$$

where $p(m_i)$ is the fractional population in the initial state of the sublevel with $J_z = m_i\hbar$, and $|\langle J_i m_i|J_f m_f LM\rangle|^2$ is the Clebsch-Gordon coefficient that describes the angular momentum piece of the overlap between the initial state and the final state of the nucleus plus radiation field. The X_L^M's are vector spherical harmonics (see Jackson, Chapter 16) that describe the angular distribution of photons emitted in a radiation field which has total angular momentum L and z-component M. From part (a) of the question, we know that only $L = 1$ is allowed in the transition we are considering, so (16.84) becomes

$$w(\theta) = a|X_1^0|^2 + b|X_1^{\pm 1}|^2 . \qquad (16.85)$$

Since for $L = J_i = 1$ and $J_f = 0$, the Clebsch-Gordon coefficients are simply δ_{Mm_i}, (16.84) gives $a = p(0)$ and $b = p(+1) + p(-1)$. Since we are given the θ dependence of this distribution, we can use the $L = 1$ vector spherical harmonics,

$$|X_1^0|^2 = \frac{3}{8\pi}(\sin^2 \theta), \qquad (16.86)$$

$$|X_1^{\pm 1}|^2 = \frac{3}{16\pi}(1 + \cos^2 \theta) , \qquad (16.87)$$

to find that $b = 4a$.

The angular distribution for the emission of a spin-zero particle will be given by (16.84), with the replacement $|X_L^M| \to |Y_L^M|$, where the Y_L^M's are spherical harmonics. Since the initial populations of the different m levels will remain the same, the angular distribution for the emission of a scalar particle will be

$$w(\theta) \propto 4|Y_1^{\pm 1}|^2 + |Y_1^0|^2 \propto 1 + \sin^2 \theta. \qquad (16.88)$$

c) The series of decays we are interested in is

$$0^+ \xrightarrow{\gamma_1} 1^+ \xrightarrow{\gamma_2} 0^+ . \qquad (16.89)$$

To find the angular correlation, we define the z quantization axis to be the direction of the first photon. Since $|X_1^0|^2$ vanishes at $\theta = 0$, the photon cannot be in a state with $M = 0$. Therefore it must have $M = \pm 1$, and by conservation of angular momentum the intermediate boron nucleus must be in a state with $m = \mp 1$. The subsequent decay to the 0^+ state will give off photons with a distribution (or an angular correlation, since $\theta_1 = 0$) described by $|X_1^{\pm 1}|^2$, or

$$w(\theta) \propto (1 + \cos^2 \theta) . \qquad (16.90)$$

Solution 7.8. a) Dysprosium is an even-even nucleus, and the ground state therefore almost certainly has all nucleons paired to give a state of total spin zero and positive parity, $J^\pi = 0^+$. Such deformed nuclei characteristically have a series of low-lying energy levels arising from collective rotations, each with angular momentum an even multiple of $\hbar$. (The states with odd angular momentum are forbidden by requiring total symmetry of the wavefunction.) The energies needed to excite rotations are typically smaller than those needed to excite vibrations or to excite nucleons to different shell model states.

Define the total angular momentum $\mathbf{J} = \mathbf{R} + \mathbf{j}$, where $\mathbf{R}$ is the rotational angular momentum of the nucleus, and $\mathbf{j}$ is its intrinsic spin. The energy of a state with collective rotational angular momentum $R = 2n$ is $E = \langle \mathbf{R}^2 \rangle / 2I - 2n(2n+1)\hbar^2/2I$, where I is the moment of inertia of the nucleus. If we use the data given in the question, we can check that $E/2n(2n+1)$ is a constant for the levels shown, as expected. All the states in the band will have the same parity.

For holmium, we do not have such a simple argument for finding the ground state. However, we are told that ^{165}Ho has $J^\pi = 7/2^-$. This isotope differs from ^{163}Ho by two neutrons. Provided these extra neutrons are paired, the two isotopes should have the same spin-parity. Therefore we suggest that the ground state of ^{163}Ho also has $J^\pi = 7/2^-$.

The low-lying states will again be rotational levels. These states have energies proportional to $\langle \mathbf{R}^2 \rangle = \langle \mathbf{J}^2 + \mathbf{j}^2 - 2\mathbf{J}\cdot\mathbf{j} \rangle$. If $K\hbar$ is the conponent of $\mathbf{j}$ on the nuclear symmetry axis, it can be shown that $\langle \mathbf{J}\cdot\mathbf{j} \rangle = K^2\hbar^2$, provided $K \neq 1/2$. Different values of R give rise to excited levels which fall into bands, each band being labeled by a different value of K, and containing states of spin $J = K + n$, for non-negative integers n.

The lowest energy band corresponds to $K = 7/2$, the maximum possible value. We can check that the excitation energies of the levels in this band are indeed given to good accuracy by the formula $E_J = [J(J+1) - K(K+1)]\hbar^2/2I$. All the states in the band will have the same parity, i.e., negative, but the parity can change between bands. (This subject is dealt with in more detail in Frauenfelder and Henley, Chapter 16.)

b) Dysprosium has no intrinsic spin, so the total angular momentum is given by the collective rotation, $\mathbf{J} = \mathbf{R}$. Furthermore, by a standard argument (e.g., see Problem 4.2) the nucleus cannot rotate about its symmetry axis (shown in Figure 16.1 as $\hat{\mathbf{z}}'$), and so the collective rotation $\mathbf{R}$ must be perpendicular to this axis.

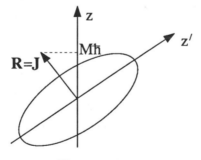

Figure 16.1.

Holmium has intrinsic spin $j = 7/2$, and to this we must add the collective angular momentum $\mathbf{R}$, to form the total angular momentum $\mathbf{J}$. Three quantities are conserved. Conventionally these are taken to be the magnitude of $\mathbf{J}$, $J\hbar$; the component of the intrinsic spin along the symmetry axis $\hat{\mathbf{z}}'$, which will have magnitude $K\hbar$; and $M\hbar$, the component of **J** as measured along the laboratory $\hat{z}$-axis. We will label states with J, K and M. The resulting vector-angular-momentum diagram is shown in Figure 16.2.

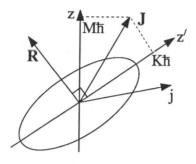

Figure 16.2.

c) To calculate the magnetic moments of the excited states of dysprosium, we approximate the nucleus as A identical nucleons, each with mass m_p and charge $q = eZ/A$, orbiting at an average radius r with speed v. The average current arising from this motion is $I = Aqv/(2\pi r)$, enclosing an area πr^2. A standard result of electrodynamics tells us that the magnetic moment μ of the current loop is

$$\mu = \frac{I}{c} \times (\text{Area}) = \frac{Aqv}{2\pi rc} \times \pi r^2 = \frac{eZvr}{2c} . \qquad (16.91)$$

We can rewrite Am_pvr as the rotational angular momentum $R\hbar$:

$$\mu = \frac{eZ}{A} \times \frac{R\hbar}{2m_pc} = \frac{Z}{A}\mu_N R , \qquad (16.92)$$

where $\mu_N = e\hbar/2m_pc$ is the nuclear magneton. For the particular case of dysprosium, $Z/A \approx 0.4$ and so we expect the moments of the states with $J = R = 2, 4$ to be around $0.8\mu_N$ and $1.6\mu_N$ respectively. These are reasonable estimates, compared with the measured values.

To calculate the radiative lifetimes, we note that the radiation is between adjacent excited states, and so from part (a) the change in angular momentum of the nucleus is $\Delta J = 2$. (Transitions with higher ΔJ are heavily suppressed.) The dominant radiation will therefore be quadrupole, i.e., $E2$ or $M2$ (because higher multipoles are supressed).

However, magnetic quadrupole is ruled out, as the selection rules for $M2$ require a change in parity. For a quadrupole decay, the rate is proportional to E^5, where E is the energy of the transition (this follows from the general formula that the rate is proportional to $E^{(2L+1)}$, for radiation of the Lth pole). Thus we expect $(\tau_{1/2})E^5$ to be constant, ignoring possible differences in the matrix elements. Using the data given, we see that this is indeed approximately true. (The agreement is especially good considering we are taking such a high power of the energy.)

d) To derive a formula for the magnetic moment of an odd-A nucleus, we proceed along the lines of the derivation of the formula for Schmidt limits. (See Problem 7.2.) We add the contributions from the collective and intrinsic angular momenta $\mathbf{R}$ and $\mathbf{j}$ to find the magnetic dipole operator,

$$\boldsymbol{\mu} = \frac{1}{\hbar}\left(g_R\mu_N\mathbf{R} + g_K\mu_N\mathbf{j}\right) . \tag{16.93}$$

Our object is to find an expression for this in terms of the conserved quantum numbers of the motion, namely J, M, and K. The expectation value of the dipole operator must be aligned with the axis of the total nuclear spin (since this is the only direction defined in the nucleus), i.e., $\langle\boldsymbol{\mu}\rangle = k\langle\mathbf{J}\rangle$. An application of the Wigner–Eckart theorem allows us to eliminate the constant k and write

$$\langle\boldsymbol{\mu}\rangle = \frac{\langle\boldsymbol{\mu}\cdot\mathbf{J}\rangle}{J(J+1)\hbar^2}\langle\mathbf{J}\rangle . \tag{16.94}$$

The magnetic moment μ is defined to be the expectation value of μ_z in a state with $J_z = J\hbar$,

$$\mu \equiv \frac{\langle\boldsymbol{\mu}\cdot\mathbf{J}\rangle}{(J+1)\hbar} . \tag{16.95}$$

If we take the scalar product of (16.93) with $\mathbf{J}$, we get

$$\langle\boldsymbol{\mu}\cdot\mathbf{J}\rangle = \frac{\mu_N}{2\hbar}\langle\{(g_R+g_K)(\mathbf{R}+\mathbf{j}) + (g_R-g_K)(\mathbf{R}-\mathbf{j})\}\cdot\mathbf{J}\rangle \tag{16.96}$$

$$= \frac{\mu_N}{2\hbar}\left\{(g_R+g_K)J(J+1)\hbar^2 + (g_R-g_K)\langle\mathbf{R}^2-\mathbf{j}^2\rangle\right\} .$$

Now we write

$$\langle\mathbf{R}^2\rangle = \langle(\mathbf{J}-\mathbf{j})^2\rangle = (J(J+1) + j(j+1) - 2K^2)\hbar^2 , \tag{16.97}$$

as stated in part (a). Together with equations (16.95) and (16.96), this allows us to arrive at the final expression for μ:

$$\mu = \mu_N J \left\{ g_R + (g_K - g_R) \frac{K^2}{J(J+1)} \right\}. \tag{16.98}$$

The collective gyromagnetic ratio, g_R, is just the quantity Z/A that we considered in part (c). One nonrigorous argument for estimating g_K is to say that the total magnetic moment arising from the nucleonic angular momentum should be about the same as that due to a single free proton ($\mu_p = 2.7\mu_N$), since the magnetic moment arose in the first place from an unpaired proton. In this case, we can write $g_K \mu_N j \approx 2.7 \mu_N$, or $g_K \approx 0.8$. Using the ground-state values $J = K = 7/2$, and $Z/A \approx 0.4$, we find from equation (16.98) a value of $\mu \approx 2.5\mu_N$.

Solution 7.9. a) The general non-relativistic nucleon operator corresponding to Fermi beta decay is

$$\mathcal{O}_F = G_V \sum_k \tau_k^{\pm}, \tag{16.99}$$

where τ_k^+ (τ_k^-) is the isospin raising (lowering) operator on the nucleon labeled by k, with the normalization convention that $\tau^+ |n\rangle = |p\rangle$, where $|n\rangle$ is a neutron and $|p\rangle$ a proton. For a single nucleon the sum collapses to a single term, of course. The decay rate for a Fermi beta decay from a state $|Nl\pi t_z jm\rangle$ to a state $|N' l' \pi' t'_z j' m'\rangle$ is proportional to the squared matrix element of $\mathcal{O}_F$ between these states:

$$\Gamma \propto \left| \langle N' l' \pi' t'_z j' m' | \tau^+ | Nl\pi t_z jm \rangle \right|^2. \tag{16.100}$$

Thus we have the following selection rules:

$$\Delta t_z = \pm 1, \text{ and}$$
$$\Delta N = \Delta l = \Delta \pi = \Delta j = \Delta m = 0. \tag{16.101}$$

The general non-relativistic nucleon operator which corresponds to Gamow-Teller beta decay is

$$\mathcal{O}_{GT} = G_A \sum_k \tau_k^{\pm} \sigma_k, \qquad (16.102)$$

where σ is the vector of Pauli spin matrices $(\sigma_x, \sigma_y, \sigma_z)$, or equivalently $(\sigma^+, \sigma^-, \sigma_z)$, with $\sigma^{\pm} = (\sigma_x \pm i\sigma_y)/\sqrt{2}$. Here we find for a single nucleon

$$\Gamma \propto \left| \langle N' l' \pi' t'_z j' m' | \tau^{\pm} \sigma | N l \pi t_z j m \rangle \right|^2, \qquad (16.103)$$

so that $\Delta N = \Delta l = \Delta \pi = 0$ and $\Delta t_z = \pm 1$ are required for nonzero Γ. The selection rules for Δj and Δm follow from the fact that σ is a vector operator and $\mathbf{j}$ is the total angular momentum, and are

$$\Delta j = 0, \pm 1 \quad \text{and} \quad \Delta m = 0, \pm 1, \qquad (16.104)$$
$$\text{but } j' = 0 \nrightarrow j = 0 \qquad .$$

These are the selection rules on j and m for any vector operator acting in angular momentum space, whether the operator is the familiar electric dipole operator, which causes the E1 trasition, or the σ operator in this problem.

b) Given that $L = 0$ for each nucleus, we need to figure out the values for the total spin S, the total isospin T, and total angular momentum J for each nucleus. For ^{6}He, $T = 1$ because the two valence nucleons are both neutrons with $t_z = -1/2$. In general, $S = 0$ or 1 for two spin 1/2 particles. Here, we are told $J = 0$ and $L = 0$ so that $S = 0$ is required. Thus the wavefunction for ^{6}He is

$$|\Psi_H\rangle = |S = 0, m_S = 0, T = 1, T_z = -1\rangle \qquad (16.105)$$
$$= \frac{1}{\sqrt{2}} [|\uparrow_1 \downarrow_2\rangle - |\downarrow_1 \uparrow_2\rangle] |n_1 n_2\rangle,$$

where $|\uparrow_1 \downarrow_2\rangle$ indicates that the spin of nucleon 1 is up and the spin of nucleon 2 is down, and where $|n_1 n_2\rangle$ indicates that nucleons 1 and 2 are both neutrons. In the case of ^{6}Li, the valence nucleons are a proton $(t_z = +1/2)$ and a neutron $(t_z = -1/2)$, so that the isospin due to the valence nucleons is $T = 0$ or 1. We know $L = 0$ and $J = 1$ so we must have $S = 1$. For overall antisymmetry of the wavefunction, we require $L + S + T = $ odd, so the minimum possible isospin is $T = 0$. This means that the three wavefunctions possible for ^{6}Li are

$$|\Psi_L(m_S)\rangle = |T = 0, T_z = 0, S = 1, m_S = \{0, \pm 1\}\rangle \qquad (16.106)$$

$$= \frac{1}{\sqrt{2}}\left\{\begin{array}{c} |\uparrow_1\uparrow_2\rangle \\ |\downarrow_1\downarrow_2\rangle \\ \frac{1}{\sqrt{2}}(|\uparrow_1\downarrow_2\rangle + |\downarrow_1\uparrow_2\rangle) \end{array}\right\} [|n_1p_2\rangle - |p_1n_2\rangle].$$

The nucleon operator governing the beta decay of ^{6}He to ^{6}Li is the Gamow-Teller operator (16.102), since $\Delta J = -1$ is Fermi-forbidden. Thus, the sum over the squared matrix elements of all possible decays is

$$\sum_i |\mathcal{M}_i|^2 = G_A^2 \sum_{m_S = 0, \pm 1} |\langle\Psi_L(m_S)| \sum_{k=1}^{2} \tau_k^+ \sigma_k |\Psi_H\rangle|^2. \qquad (16.107)$$

We will first consider $\sum_k \tau_k^+ \sigma_k |\Psi_H\rangle$. Recall that the components of σ are

$$\sigma_x = \begin{pmatrix} 0 & 1 \\ 1 & 0 \end{pmatrix}, \quad \sigma_y = \begin{pmatrix} 0 & -i \\ i & 0 \end{pmatrix} \quad \text{and } \sigma_z = \begin{pmatrix} 1 & 0 \\ 0 & -1 \end{pmatrix}. \qquad (16.108)$$

Note that the up and down spin eigenvectors are $\uparrow = \begin{pmatrix} 1 \\ 0 \end{pmatrix}$ and $\downarrow = \begin{pmatrix} 0 \\ 1 \end{pmatrix}$ in this notation. We also have

$$\sigma^+ = \begin{pmatrix} 0 & \sqrt{2} \\ 0 & 0 \end{pmatrix} \quad \text{and } \sigma^- = \begin{pmatrix} 0 & 0 \\ \sqrt{2} & 0 \end{pmatrix}, \qquad (16.109)$$

where $\sigma^\pm = (\sigma_x \pm i\sigma_y)/\sqrt{2}$ as in (a). Using these definitions, we find the following:

$$\begin{array}{ll} \sigma^+|\uparrow\rangle = 0, & \sigma^+|\downarrow\rangle = \sqrt{2}|\uparrow\rangle, \\ \sigma^-|\uparrow\rangle = \sqrt{2}|\downarrow\rangle, \quad \text{and} & \sigma^-|\downarrow\rangle = 0. \end{array} \qquad (16.110)$$

(Notice that $\sigma^\pm$ differ in their normalizations from the isospin operators, which are defined so that $\tau^+|n\rangle = a|p\rangle$), where $a = 1$ rather than $a = \sqrt{2}$.) We are now ready to evaluate $\sum_k \tau_k^+ \sigma_k |\Psi_H\rangle$. Let us consider each component of $\boldsymbol{\sigma}$ separately. First,

$$\sum_{k=1}^{2} \tau_k^{+} \sigma_k^{+} |\Psi_H\rangle = \frac{1}{\sqrt{2}} \left[\sigma_1^{+}| \uparrow_1\downarrow_2\rangle - \sigma_1^{+}| \downarrow_1\uparrow_2\rangle \right] \tau_1^{+}|n_1 n_2\rangle$$

$$+ \frac{1}{\sqrt{2}} \left[\sigma_2^{+}| \uparrow_1\downarrow_2\rangle - \sigma_2^{+}| \downarrow_1\uparrow_2\rangle \right] \tau_2^{+}|n_1 n_2\rangle$$

$$= \frac{1}{\sqrt{2}} \left[-| \uparrow_1\uparrow_2\rangle|p_1 n_2\rangle + | \uparrow_1\uparrow_2\rangle|n_1 p_2\rangle \right]$$

$$= -\sqrt{2}|S = 1, m_S = 1, T = 0, T_z = 0\rangle. \quad (16.111)$$

Similarly, we find

$$\sum_{k=1}^{2} \tau_k^{+} \sigma_k^{-} |\Psi_H\rangle = \sqrt{2}|S = 1, m_S = -1, T = 0, T_z = 0\rangle, \quad (16.112)$$

and finally,

$$\sum_{k=1}^{2} \tau_k^{+} (\sigma_z)_k |\Psi_H\rangle = \sqrt{2}|S = 1, m_S = 0, T = 0, T_z = 0\rangle. \quad (16.113)$$

It is now simple to sum the squared matrix elements since each of the components above corresponds to one of the three possible ^{6}Li wavefunctions given in equation (16.106):

$$\sum |\mathcal{M}_i|^2 = 3(\sqrt{2})^2 G_A^2 = 6G_A^2. \quad (16.114)$$

c) The idea here is that

$$\frac{1}{f\tau} = \mathcal{K} \sum_j |\mathcal{M}_j|^2, \quad (16.115)$$

where $\mathcal{K}$ is a constant common to both neutron decay and the ^{6}He beta decay. We take the initial wavefunction of the neutron as $| \uparrow n\rangle$ (arbitrarily choosing spin up). The neutron can decay to both the states $| \uparrow p\rangle$ and $| \downarrow p\rangle$. Thus neutron decay involves *both* a Fermi component *and* a Gamow-Teller component. The sum of squared matrix elements is

$$\sum |\mathcal{M}_i|^2 = G_V^2 |\langle \uparrow p|\tau^{+}| \uparrow n\rangle|^2 +$$

$$G_A^2 |\langle \uparrow p|\tau^{+}\sigma_z| \uparrow n\rangle|^2 + G_A^2 |\langle \downarrow p|\tau^{+}\sigma^{-}| \uparrow n\rangle|^2$$

$$= G_V^2 + 3G_A^2 = G_A^2(G_V^2/G_A^2 + 3). \quad (16.116)$$

Using equations (16.114) and (16.115) we write

$$\frac{(f\tau)_{\text{He}}}{(f\tau)_{\text{n}}} = \frac{3 + G_V^2/G_A^2}{6}, \tag{16.117}$$

and using the information given, we have $(f\tau)_{\text{He}} \approx 680$ s, which is within twenty percent of the experimental value of about 810 seconds.

Solution 7.10. a) Let us denote the 20 MeV excited state in ^{4}He by ^{4}He*. Since the rest mass energy in (^{3}He + n) is only a couple of widths higher than the energy of the resonance (i.e., the ^{4}He* state), the reaction ^{3}He + n → $t + p$ will proceed by the formation of an intermediate ^{4}He* virtual state, at least at low neutron kinetic energy. However, this virtual state can only be formed if the quantum numbers of the ($n + ^3$He) state agree with the quantum numbers of the ^{4}He* state.

The two incoming particles are both spin 1/2, and so the total spin S is either zero or one. Since the spin of the ^{4}He* is zero, the ($n + ^3$He) system must have total spin zero for the reaction to ^{4}He* to proceed. If the spins of the ^{3}He and n are parallel then they must be in a spin 1 state, so that no neutrons will be absorbed by the dominant process of ^{3}He + n →^{4} He*. If, however, the spins of the particles are antiparallel, then there is a 50% probability that the particles are in a spin 0 state. Therefore, if the ^{3}He is polarized in the direction of the neutron momentum, only neutrons with helicity of -1 will be absorbed by the process ^{3}He + n →^{4} He* → $t + p$. Of course, other processes (assumed to be much weaker) may absorb or scatter neutrons of either helicity.

b) By Fermi's golden rule, the rate for ^{3}He + n → $t + p$ is given by

$$\Gamma = \frac{2\pi}{\hbar} |M|^2 \rho(E), \tag{16.118}$$

where M is the matrix element describing the process, and $\rho(E)$ is the phase space available to the outgoing particles, which have a total kinetic energy E. The energy E is the difference in rest mass of the ingoing and outgoing particles, added to the neutron's kinetic energy. In the limit that the neutron's kinetic energy is much less than the rest mass difference, $\rho(E)$ is independent of the neutron kinetic energy. We will assume that we are far enough from the peak of the resonance that the matrix element M will not be strongly energy-dependent (but close enough to the peak that the resonance still dominates the reaction). Therefore, the rate Γ is approximately independent of the neutron kinetic energy. The cross-section is the rate divided by the flux of neutrons, which is proportional to the neutron velocity v. Since we have just shown that the rate does not depend on the neutron velocity, we have the simple proportionality that the cross-section varies as $1/v$. One can also arrive at this result in an even more heuristic manner by stating that the cross-section must be proportional to the "time the neutron spends near the ^{3}He."

From the preceding argument it is clear that the relation $\sigma \propto 1/v$ is valid for any exothermic reaction when the kinetic energies of the incoming particles are low compared to the energy released by the reaction.

c) Let us define "spin up" to mean that the spin lies parallel to the beam axis, and "spin down" to mean that the spin lies antiparallel. If $\sigma_{\uparrow\uparrow}$ is the cross-section for the case when the ^{3}He and n spins are parallel, and $\sigma_{\uparrow\downarrow}$ is the cross-section for the case when the two spins are antiparallel, then an unpolarized beam of neutrons is absorbed with cross-section $\sigma = (\sigma_{\uparrow\uparrow} + \sigma_{\uparrow\downarrow})/2$. Following the arguments of part (a), we will assume that $\sigma_{\uparrow\uparrow} \approx 0$, so that $\sigma_{\uparrow\downarrow} \approx 2\sigma = 1700$ barns.

Assume that the ^{3}He target is of uniform thickness. Let $n_\uparrow$ be the number of ^{3}He with spin up per unit length and $n_\downarrow$ be the number with spin down per unit length. Similarly let $m_\uparrow$ and $m_\downarrow$ be the fluxes of neutrons with spin up and spin down respectively. Then we can write down the differential equations governing $m_\uparrow$ and $m_\downarrow$ as a function of x, the distance along the beam axis inside the ^{3}He target:

$$\frac{dm_\downarrow}{dx} = -2\sigma n_\uparrow m_\downarrow \quad \text{and} \quad \frac{dm_\uparrow}{dx} = -2\sigma n_\downarrow m_\uparrow, \qquad (16.119)$$

where we have used the assumption than $\sigma_{\uparrow\downarrow} \approx 2\sigma$ and $\sigma_{\uparrow\uparrow} \approx 0$. For a target of thickness t, the flux of the neutrons downstream from the target is given by

$$m_\downarrow = m_0 e^{-2\sigma n_\uparrow t} \quad , \quad m_\uparrow = m_0 e^{-2\sigma n_\downarrow t}, \tag{16.120}$$

where m_0 is half the incident flux on the target. The polarization of the neutrons is

$$P = \frac{m_\uparrow - m_\downarrow}{m_\uparrow + m_\downarrow} = \frac{\exp(-2\sigma n_\downarrow t) - \exp(-2\sigma n_\uparrow t)}{\exp(-2\sigma n_\downarrow t) + \exp(-2\sigma n_\uparrow t)}. \tag{16.121}$$

From the given information that $(n_\uparrow + n_\downarrow)t = 6 \times 10^{21}$ atoms/cm^2 and $(n_\uparrow - n_\downarrow)t = 0.65(n_\uparrow + n_\downarrow)t$, we find that $n_\uparrow t = 4.95 \times 10^{21}$ atoms/cm^2 and $n_\downarrow t = 1.05 \times 10^{21}$ atoms/cm^2. Evaluating equation (16.121) yields a value of 99.8% for the polarization of the emerging neutrons.

Chapter 17

Elementary Particle Physics—Solutions

Solution 8.1. a) The decays $K^0 \rightarrow \pi^+\pi^-$ and $K^0 \rightarrow \pi^0\pi^0$ must be weak decays, since the strangeness changes by one. Because the pions are isospin 1 particles, they combine to form states with a total isospin, I, of zero, one or two. We need to know the symmetries of these states. We can quote the general result that when we add two spins (or isospins) of value j, the total angular momentum can take all integer values from $2j$ to zero: the states $2j$, $2j - 2,\ldots$ are symmetric under particle interchange, while the states $2j - 1$, $2j - 3,\ldots$ are antisymmetric. Applying these considerations to the present problem, we see that the $I = 2$ and $I = 0$ states are symmetric and the $I = 1$ state is antisymmetric.

The "$\Delta I = 1/2$" rule states that, for strangeness-changing decays, processes in which the total isospin changes by $1/2$ are favored. (Often a fictitious isospin $1/2$ particle called a spurion is invoked to apply this rule.) Since the isospin of the kaon is $1/2$, the total isospin of the two pions is most likely 0 or 1. The kaon has zero total angular momentum in its rest frame. The pions are spinless, so they must also have zero orbital angular momentum, and therefore their spatial wavefunction is symmetric. Since they are bosons, the total wavefunction must be

256

symmetric. Therefore, the total isospin must be zero.

b) The probability of decay is proportional to a matrix element whose isospin part is given by the Clebsch-Gordon coefficients $|\langle \pi^0 \pi^0 | I = 0 \ I_z = 0 \rangle|^2$ for $K \to \pi^0 \pi^0$, and $|\langle \pi^+ \pi^- | I = 0 \ I_z = 0 \rangle|^2$ for $K \to \pi^+ \pi^-$. If we label states in terms of the isospin $I^{(i)}$ of each individual particle as $| \ I^{(1)} \ I^{(2)} \ I_z^{(1)} \ I_z^{(2)} \rangle$ then

$$| \pi^0 \pi^0 \rangle = | 1 \ 1 \ 0 \ 0 \rangle \text{ and} \tag{17.1}$$

$$| \pi^+ \pi^- \rangle = \frac{1}{\sqrt{2}} \left(| 1 \ 1 \ +1 \ -1 \rangle + | 1 \ 1 \ -1 \ +1 \rangle \right). \tag{17.2}$$

where we have necessarily written the $| \pi^+ \pi^- \rangle$ as a symmetric combination, since they are bosons and the rest of the wavefunction is symmetric (see part (a)). The $\pi^0 \pi^0$ and $\pi^+ \pi^-$ states differ only in their isospin part, so

$$\frac{\Gamma(K \to \pi^+ \pi^-)}{\Gamma(K \to \pi^0 \pi^0)} = \frac{\frac{1}{2} |\langle 1 \ 1 \ 1 \ -1 | 0 \ 0 \rangle + \langle 1 \ 1 \ -1 \ 1 | 0 \ 0 \rangle|^2}{|\langle 1 \ 1 \ 0 \ 0 | 0 \ 0 \rangle|^2} = 2. \tag{17.3}$$

Somewhat hidden in the isospin formalism above is that the factor of two is due to the fact that the neutral pions are identical particles, and the π^+ and π^- are not.

c) The mass of three pions is approximately 3×140 MeV$=420$ MeV, which is close to the mass of the kaon. Therefore the difference in decay rates is largely due to differing phase-space factors between the two- and three-pion decays.

d) The CP eigenstates are

$$| K_1 \rangle = \frac{1}{\sqrt{2}} \left(| K^0 \rangle + | \overline{K^0} \rangle \right),$$

$$| K_2 \rangle = \frac{1}{\sqrt{2}} \left(| K^0 \rangle - | \overline{K^0} \rangle \right), \tag{17.4}$$

where it is clear that $CP| K_1 \rangle = | K_1 \rangle$, and $CP| K_2 \rangle = - | K_2 \rangle$.

e) The energy eigenstates of the neutral kaon are $|K_1\rangle$ and $|K_2\rangle$. Therefore the wavefunction for an arbitrary superposition of neutral kaons as a function of proper time is

$$| \Psi(t) \rangle = c_1 | K_1 \rangle e^{-im_1 t} e^{-\Gamma_1 t/2} + c_2 | K_2 \rangle e^{-im_2 t} e^{-\Gamma_2 t/2}, \tag{17.5}$$

where m_1 and m_2 are the masses of K_1 and K_2. The initial condition that $|\Psi(0)\rangle = |K^0\rangle = (|K_1\rangle + |K_2\rangle)/\sqrt{2}$ demands that the constants c_1 and c_2 have the values $c_1 = c_2 = 1/\sqrt{2}$. The ratio at time t of $\overline{K^0}$ to K^0 is

$$\frac{|\langle \Psi(t) | \overline{K^0}\rangle|^2}{|\langle \Psi(t) | K^0\rangle|^2} = \frac{\left| e^{-im_1 t}e^{-\Gamma_1 t/2} - e^{-im_2 t}e^{-\Gamma_2 t/2} \right|^2}{\left| e^{-im_1 t}e^{-\Gamma_1 t/2} + e^{-im_2 t}e^{-\Gamma_2 t/2} \right|^2}$$

$$= \frac{e^{-\Gamma_1 t} + e^{-\Gamma_2 t} - 2e^{-\Gamma t}\cos \Delta mt}{e^{-\Gamma_1 t} + e^{-\Gamma_2 t} + 2e^{-\Gamma t}\cos \Delta mt}, \qquad (17.6)$$

where $\Gamma = (\Gamma_1 + \Gamma_2)/2$ and $\Delta m = m_1 - m_2$.

Solution 8.2. a) The square of the four-momentum transfered to the target is

$$q^2 = (k_i - k_f)^2, \qquad (17.7)$$

where k_i is the initial four-momentum of the target quark, and k_f is its final four-momentum. In the initial rest frame of the quark, these four-momenta are

$$k_i = \begin{pmatrix} m_q \\ 0 \\ 0 \\ 0 \end{pmatrix} \text{ and } k_f = \begin{pmatrix} m_q + \nu \\ p \\ 0 \\ 0 \end{pmatrix}, \qquad (17.8)$$

where m_q is the quark mass. We use these expressions in equation (17.7) to find

$$q^2 = \nu^2 - p^2 = \nu^2 + m_q^2 - E_q^2, \qquad (17.9)$$

where $E_q = m_q + \nu$ is the final energy of the quark, and we have used the relation $E_q^2 = p^2 + m_q^2$. Substituting for E_q gives

$$q^2 = -2m_q \nu = -2x m_N \nu, \qquad (17.10)$$

which we rearrange to find x in terms of the nucleon mass m_N:

$$x = \frac{-q^2}{2m_N \nu}. \qquad (17.11)$$

b) We expect incoherent scattering from the quarks, so that we may sum the contributions to the cross-section from each type of quark present in the target. Since the scattering will be dominated by photon exchange, the scattering from a given quark flavor is proportional to the squared charge of that quark. Therefore a proton, which comprises two u quarks and one d quark, has a cross-section for scattering a muon of

$$\sigma_{\mu-p} \propto \sum_i Q_i^2 = Q_u^2 + Q_u^2 + Q_d^2 = 2\left(\frac{2}{3}\right)^2 + \left(\frac{-1}{3}\right)^2 = 1, \quad (17.12)$$

where Q_i is the charge of a quark of type i. On the other hand, the deuteron consists of three u quarks and three d quarks, so that for scattering from deuterons,

$$\sigma_{\mu-d} \propto 3Q_u^2 + 3Q_d^2 = \frac{15}{9}. \quad (17.13)$$

We therefore expect the ratio of the cross-sections to be

$$\frac{\sigma_{\mu-p}}{\sigma_{\mu-d}} \approx \frac{9}{15}. \quad (17.14)$$

c) Whenever two particles interact via exchange of a photon, there is another change that can take place: exchange of a Z boson. The amplitude for such a process is suppressed at low q^2, due to the large mass of the Z which appears in the propagator, but current experiments are sensitive enough to detect effects this small.

The correct procedure for calculating this contribution is to add coherently the amplitudes for photon and Z exchange, before squaring to find the probability. The leading correction to the purely electromagnetic process is the term arising from the interference of the two amplitudes. The calculation is straightforward but tedious. For sake of definiteness we may assume that the muon beams are polarized (μ^- left-handed and μ^+ right-handed), but this is not actually necessary.

The important result is that one of the terms arising from the interference has a sign that depends on whether the incoming lepton is a μ^-

or a μ^+, and consequently the respective scattering cross-sections are slightly different. The effect grows because it is suppressed by a factor of $1/(M_Z^2 - q^2)$, which gets bigger as q^2 increases, and clearly the scale for the growth is set by M_Z^2.

d) The rest of the momentum is carried by gluons, which only interact through the strong force. Muons are leptons, which do not interact by the strong force.

Solution 8.3.   a) Let us assume for now that the low-lying states are nonrelativistic, which will allow us to factor the wavefunctions into space and spin components. We will see in part (b) that this is justified.

Because the intrinsic spin of the scalar quarks is zero, the total spin J of the bound state will be determined by the angular momentum L. The scalar quarks are bosons, so a scalar quark σ will have the same intrinsic parity η_σ as its antiparticle: $\eta_\sigma \eta_{\bar\sigma} = +1$. The parity of the bound state will be $P = (-1)^L \eta_\sigma \eta_{\bar\sigma} = (-1)^L$. The C-parity (the eigenvalue of the charge conjugation operator) is $(-1)^{L+S} = (-1)^L$ as well. Thus the first few states have $J^{PC} = 0^{++}$, 1^{--}, 2^{++}, etc.

To determine which of these states can be produced, we can use a simple argument which is generally applicable to electromagnetic annihilations. If we are only concerned with leading-order processes, then the dominant reaction will be $e^+ e^-$ annihilation into a single photon, which then converts into a scalar quark bound state. The quantum numbers of this state must be the same as those of the intermediate photon: $J^{PC} = 1^{--}$. The only way that other states can be produced is by higher order processes, such as radiative decays of the 1^{--} states. Note that for reactions which proceed via the strong force, like those in $p\bar p$ colliders, this argument does not apply.

b) We will now show that the assumption that the bound states are nonrelativistic is self-consistent. We can write down a Schrödinger equation for the wavefunction in center-of-mass coordinates (in units where $\hbar = c = 1$):

$$-\frac{1}{2\mu}\nabla^2\psi + ar\psi = E\psi. \tag{17.15}$$

Here μ is the reduced mass, $m_\sigma/2 = 2.5$ GeV, but since we are dropping constants of order unity we will not emphasize this distinction. We could now use a standard method such as WKB to find the ground-state energy. However a much simpler way to get almost the same result is to make some crude approximations, which work well in the case of the hydrogen atom, for example, and which effectively amount to dimensional analysis. We argue (e.g., via the virial theorem) that the kinetic and potential terms will be about the same. Further, if the bound state has mean radius r_0, the magnitude of the kinetic term will be, on dimensional grounds, $1/(m_\sigma r_0^2)$, which we set equal to the potential term ar_0. Thus we arrive at the estimate $r_0 \approx (m_\sigma a)^{-1/3} \approx 10^{-3}$ MeV^{-1}. The total energy E, the sum of the kinetic and potential energies, will also be of the same size,

$$E \approx \frac{a}{(m_\sigma a)^{1/3}} \approx 170 \text{ MeV}. \tag{17.16}$$

Now we can see that the kinetic energy of each quark is much less than its mass, and relativistic corrections will be small, on the order of a few percent, provided we are only considering low excitation levels. In contrast, for a system with a $1/r$ potential, it is the low-lying states that are the *most* relativistic.

Next consider the splitting of the states. The level diagram will be very different from that of, say, positronium. Because the scalar quarks are spinless, they are forbidden from having any magnetic dipole moment. Consequently, there can be no "fine" or "hyperfine" structure. Let us consider the splitting between states with the same radial quantum number, but differing in angular momentum, for example the ground state and the state with one quantum of angular momentum. We can easily estimate the magnitude of the kinetic energy of rotation of the latter. The state has mean radius r_0 and angular momentum $L = m_\sigma r_0^2 \omega = \hbar$ (which in our units is just 1). Classically the kinetic energy is $T = m_\sigma r_0^2 \omega^2/2$, and we can eliminate ω to get $T = 1/(2m_\sigma r_0^2)$.

Thus the splitting is of order

$$\Delta E \approx \frac{1}{m_\sigma r_0^2} \approx \left(\frac{a^2}{m_\sigma}\right)^{1/3} \approx 170 \text{ MeV} . \tag{17.17}$$

Actually, we see that the splitting between the levels is of the same order as the ground state energy, and this is not surprising since we are simply using the same dimensional constants to form a quantity with the dimensions of energy. (In fact, in the case of a $1/r$ potential the splitting between states with different angular momenta is exactly equal to the splitting between states of different radial quantum number, leading to the degeneracy of the $2s$ and $2p$ states in hydrogen, for example. NB: The principle quantum number n is *not* the radial quantum number.)

To estimate $|\psi(0)|^2$, we can make the simple approximation that the bound state will be more or less uniformly distributed over a volume of order r_0^3. This gives us

$$|\psi(0)|^2 \approx \frac{1}{r_0^3} \approx m_\sigma a . \tag{17.18}$$

In the case of the Schrödinger model of hydrogen, this answer is correct up to a factor of π.

c) We will write down the gauge-invariant Lagrangian for this system term by term. The quarks are charged scalar particles, and they will be represented by a complex scalar wavefunction. The only choice for the kinetic term is $(D_\mu\phi)^\dagger (D^\mu\phi)$, where $D_\mu = \partial_\mu + iqA_\mu$ is the covariant derivative. To this we add the mass term $m_\sigma^2\phi^\dagger\phi$, and the photon kinetic term $-1/4F_{\mu\nu} F^{\mu\nu}$ to get

$$\mathcal{L} = (D_\mu\phi)^\dagger (D^\mu\phi) + m_\sigma^2\phi^\dagger\phi - \frac{1}{4}F_{\mu\nu} F^{\mu\nu} - V_{int} , \tag{17.19}$$

where V_{int} describes the color interactions between the scalar quarks. If we expand the kinetic term, we find that there are two types of vertices in this theory: $-iq(\phi A_\mu\partial^\mu\phi - \phi\partial_\mu\phi A^\mu)$, shown in Figure 17.1a, and $q^2A_\mu A^\mu\phi^\dagger\phi$, shown in Figure 17.1b.

Conservation of energy and momentum forbid decay into a single photon. There are two diagrams for the decay into two photons. The

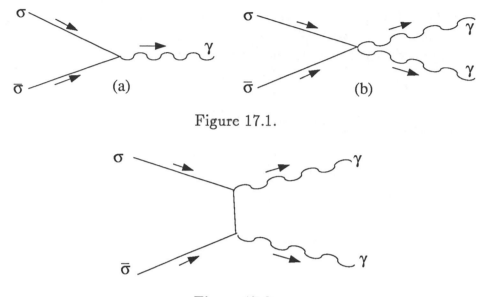

Figure 17.1.

Figure 17.2.

first is just Figure 17.1b. The second is obtained by using two vertices of the type 17.1a, as shown in Figure 17.2. Both these diagrams are of the same magnitude.

To estimate the decay width, we use an argument that closely parallels the treatment of the decay of positronium in Sakurai. First consider the annihilation cross-section for a scalar quark incident on an antiquark. At low energies this will be approximately R^2/v, where R is the "classical radius" of the scalar quark and v is the incident velocity. By analogy with the classical radius of the electron, we can write $R = \alpha/9m_\sigma$, where the factor of $1/9$ arises since the scalar quark has charge $e/3$. (The $1/v$ dependence frequently arises for low-energy exothermic reactions, e.g., see Problem 7.10.)

The annihilation rate will now be the cross-section multiplied by the "flux," which is just the velocity of one quark multiplied by the density at the origin, or simply $v\,|\psi(0)|^2$. Thus the decay rate is

$$\Gamma \approx \frac{R^2}{v} \times v \times |\psi(0)|^2 = \frac{\alpha^2 a}{81m_\sigma} \approx 2 \times 10^{-5} \text{ MeV.} \qquad (17.20)$$

This corresponds to a lifetime of around 3×10^{-17} s.

d) Over the last twenty years, a lot of work has gone into looking for particles with masses of a few GeV. In particular, any particle that is charged will couple electromagnetically, and should be produced in an electron-positron collider. (This is how the charm quark and the tau lepton were discovered.) Furthermore, the order of magnitude of the production cross-section is determined largely by the strength of electromagnetic reactions, and should be approximately the same for a bound state of scalar quarks as for a system like charmonium. Therefore, scalar quarks with masses in this range would have been detected a long time ago.

Solution 8.4. a) In order to treat the decay of the D^0, we must assume that it is well-described as a weakly-bound state of a $\bar{u}c$ quark pair. This allows us to write down the unique lowest-order Feynman diagram for the process $D^0 \rightarrow \mu^+ \nu X$, Figure 17.3a. Note that after the decay, the

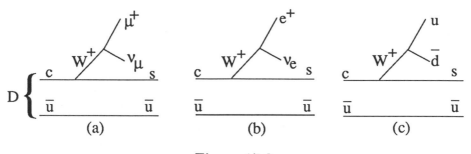

Figure 17.3.

s and $\bar{u}$ will in general be in some excited state which subsequently decays into a complicated mixture of hadrons, which we denote by X. (Note that the decay $c \rightarrow d$ happens much less frequently.)

Competing with this process are the decays $D \rightarrow e^+ \nu X$ and $D \rightarrow u\bar{d}X$, shown in Figures 17.3b,c. In the latter case, the $u\bar{d}$ quarks will also hadronize into a shower of particles, Y, but we assume that this does not affect the amplitude for this diagram. To estimate the branching ratio for the decay to muons we note that, apart from two complications, the contributions from each of these diagrams are equal. The first complication is that the phase-space factors for the final states are different; we can ignore this as long as the masses of the particles in the final states are much less than the mass of the D^0, and indeed this is true. The second complication, which we do not ignore, is that the $u\bar{d}$ quarks in Figure (17.3c) can be in any of three color states, which we have to count separately. Thus the decay to quarks occurs three times as often as either of the other two decays. In this way we see that the branching ratio for $D^0 \rightarrow \mu^+\nu X$ is approximately 1/5.

The decay rate of the D^0 is best estimated by comparison with the decay of the muon (Figure 17.4), which is the prototype for all weak decays. If we ignore the complications of the "spectator" quark $\bar{u}$ in

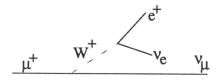

Figure 17.4.

Figure (17.3a), then these two decays are analogous, up to the differing phase-space factors. Thus we can avoid all the complications of calculating Feynman diagrams, remembering the weak coupling constant, and keeping track of factors like $192\pi^3$. Instead, we simply have to remember useful quantities like the muon lifetime, $\tau_\mu \approx 2 \times 10^{-6}$s, its mass, $m_\mu \approx 100$ MeV, and the energy dependence of phase space in three-body decays, $\rho(E) \propto E^5$.

We can now write

$$1/\tau_\mu = K \times (m_\mu)^5, \tag{17.21}$$

where K contains all the information about the Feynman diagram and the coupling constants of the theory. Thus the decay rate of the D^0 is

$$1/\tau_D \approx 5K \times (m_D)^5. \tag{17.22}$$

The factor of five counts the number of distinct decay modes, as explained above, and m_D is the energy of the final state, if we make the approximation that all the other particles are massless. Thus we find

$$\tau_D \approx \frac{1}{5} \left(\frac{m_\mu}{m_D}\right)^5 \tau_\mu \approx 2 \times 10^{-13} \text{ s.} \qquad (17.23)$$

This agrees well with the experimental value of 4×10^{-13}s.

b) A feature of systems that mix is that the mass eigenstates are not the same as the decay eigenstates. However, it is the D^0 and $\bar{D}^0$ states which decay. In particular, the D^0 can decay to a μ^+, and the $\bar{D}^0$ to a μ^-.

We may write the wavefunctions for D_1 and D_2 in terms of $|D^0\rangle$ and $|\bar{D}^0\rangle$:

$$|D_1\rangle = \frac{1}{\sqrt{2}} \left(|D^0\rangle + |\bar{D}^0\rangle\right), \ |D_2\rangle = \frac{1}{\sqrt{2}} \left(|D^0\rangle - |\bar{D}^0\rangle\right). \qquad (17.24)$$

Let us drop the cumbersome ket notation and write the time dependence of the wavefunction as

$$D_k(t) = D_k e^{-im_k t - \Gamma_k t/2}, \quad k = 1, 2 \ , \qquad (17.25)$$

where m_k is the mass of the particle and Γ_k is its decay rate.

Suppose we start off with a pure D^0 state. Then at $t = 0$ we can invert (17.24) and write the wavefunction ψ of the D as

$$\psi(t = 0) = D^0 = \frac{1}{\sqrt{2}} (D_1 + D_2). \qquad (17.26)$$

At time t we will have

$$
\begin{aligned}
\psi(t) &= \frac{1}{\sqrt{2}} [D_1(t) + D_2(t)] \\
&= \frac{1}{2} \left[\left(D^0 + \bar{D}^0\right) e^{-im_1 t - \Gamma_1 t/2} + \left(D^0 - \bar{D}^0\right) e^{-im_2 t - \Gamma_2 t/2}\right] \\
&= \frac{1}{2} \left[D^0 \left(e^{-im_1 t - \Gamma_1 t/2} + e^{-im_2 t - \Gamma_2 t/2}\right) + \right. \\
&\qquad \left. \bar{D}^0 \left(e^{-im_1 t - \Gamma_1 t/2} - e^{-im_2 t - \Gamma_2 t/2}\right)\right]. \qquad (17.27)
\end{aligned}
$$

From this expression we can pick out the probabilities for the system to be in a D^0 or a $\bar{D}^0$ state at time t, namely the absolute squares of the appropriate coefficients in the last equation.

The probability that the meson decays to a μ^+ is proportional to the probability that the meson is a D^0. Similarly, the probability that the meson decays to a μ^- is proportional to the probability that it is a $\bar{D}^0$. The constant of proportionality is the same for both decays. Therefore the ratio $R = \text{Prob}(\mu^-)/\text{Prob}(\mu^+)$ is given by

$$R = \frac{\int P_{\bar{D}}(t)\, dt}{\int P_D(t)\, dt}, \qquad (17.28)$$

where the probability of having a D^0 is

$$P_D(t) = \frac{1}{4}\left[e^{-\Gamma_1 t} + e^{-\Gamma_2 t} + 2e^{-(\Gamma_1+\Gamma_2)t/2}\cos(m_1 - m_2)t\right], \qquad (17.29)$$

and the probability of having a $\bar{D}^0$ is

$$P_{\bar{D}}(t) = \frac{1}{4}\left[e^{-\Gamma_1 t} + e^{-\Gamma_2 t} - 2e^{-(\Gamma_1+\Gamma_2)t/2}\cos(m_1 - m_2)t\right]. \qquad (17.30)$$

We integrate these probabilities from $t = 0$ to $t = \infty$ to get our final answer for the ratio:

$$R = \frac{(\Delta m)^2 + (\Delta\Gamma)^2}{2\Gamma_0^2 + (\Delta m)^2 - (\Delta\Gamma)^2}. \qquad (17.31)$$

Here we have written $\Delta m = m_1 - m_2$, $\Delta\Gamma = (\Gamma_1 - \Gamma_2)/2$ and $\Gamma_0 = (\Gamma_1 + \Gamma_2)/2$. We expect Δm to be very small, as in the kaon system, where $\Delta m_K / m_K \approx 10^{-14}$. However, because the decay products of the D^0 and $\bar{D}^0$ are similar, we expect $\Gamma_1 \approx \Gamma_2$, so that $\Delta\Gamma \ll \Gamma_0$. Hence the ratio R will be very small. In the kaon system, the dominant decay for K_S is $K_S \to 2\pi$, which is CP-suppressed for K_L. Thus the decay products of K_S and K_L are very different, and $\Delta\Gamma/\Gamma_0 \approx 1$.

Solution 8.5. a) The Lagrangian is written in the form of a sum over all the fermions,

$$\mathcal{L} = i\frac{g}{2\cos\theta_W}\left\{\sum_f Z^\mu \bar{f}\gamma_\mu \frac{1}{2}(c_V^f - c_A^f \gamma^5)f\right\}, \qquad (17.32)$$

where the c_V's and c_A's are the vector and axial coupling constants for each fermion, and their values can be found by comparing the Lagrangian in the two forms (17.32) and (8.2).

Let us set $\hbar = c = 1$. The general formula for the decay of a particle at rest into two decay products is given in equation (17.85). Using this formula and noting that neutrinos are massless, we can write the differential rate per unit solid angle for the decay of the Z^0 into a given species of neutrino as

$$\frac{d\Gamma_{\nu\bar{\nu}}}{d\Omega} = \frac{|\mathcal{M}|^2_{\nu\bar{\nu}}}{64\pi^2 M_Z} ,\tag{17.33}$$

where $|\mathcal{M}|_{\nu\bar{\nu}}$ is the matrix element for the decay and M_Z is the mass of the Z^0.

In order to avoid the full-blown spinor calculation, we will write down by inspection an approximate expression for $|\mathcal{M}|^2_{\nu\bar{\nu}}$. In doing so, we lose the angular dependence of the matrix element, but the width we calculate will only be off by a small factor.

The Lagrangian (17.32) gives a squared vertex factor of $(g/2\cos\theta_W)^2$ $[2(c_V^2 + c_A^2)]$. Aside from the cancellation of the $c_V \times c_A$ term and the factor of two, this factor comes directly from squaring the coefficient of the appropriate spinors in the Lagrangian. The squared matrix element is approximately the product of the squared vertex factor and E_f^2, where $E_f = (M_Z/2)$ is the fermion energy:

$$|\mathcal{M}|^2_{\nu\bar{\nu}} \approx \left(\frac{g}{2\cos\theta_W}\right)^2 2[(c_V^\nu)^2 + (c_A^\nu)^2]E_f^2 .\tag{17.34}$$

The factor of E_f^2 must be present so that the matrix element has the correct units. Since we are ignoring any angular dependence in $|\mathcal{M}|^2_{\nu\bar{\nu}}$, the integration over solid angle gives 4π, and the decay width is

$$\begin{aligned}\Gamma_{\nu\bar{\nu}} &\approx 4\pi \left(\frac{g}{2\cos\theta_W}\right)^2 \frac{1}{64\pi^2 M_Z}\left(\frac{M_Z}{2}\right)^2 ,\\ &\approx \frac{g^2 M_Z}{256\pi\cos^2\theta_W} .\end{aligned}\tag{17.35}$$

(A more detailed calculation shows that our answer is a factor of 3/8 too small.) Substituting $g^2 = (8G_F M_W^2)/\sqrt{2}$ (where $G_F \approx 1.2 \times$

10^{-5} GeV^{-2}) and $M_W^2 = M_Z^2 \cos^2 \theta_W$ gives

$$\Gamma_{\nu\bar{\nu}} = \frac{G_F M_Z^3}{32\sqrt{2}\pi} \approx 60 \text{ MeV} . \tag{17.36}$$

To find the width for decay into any other fermion, we substitute the appropriate values for c_V and c_A into the squared matrix element (17.34).

The total width is obtained by summing over all relevant fermions, including the two other generations not explicitly included in the Lagrangian, except that we will exclude the top quark as being too heavy to be a decay product of the Z^0. We must also remember to include the threefold multiplicity of quarks due to color. The sum of the contributions of all the fermions gives the total width, $\Gamma \approx 15 \, \Gamma_{\nu\bar{\nu}} \approx 900 \text{ MeV}$. The factor of 8/3 we were missing in our calculation of $\Gamma_{\nu\bar{\nu}}$ obviously affects our value for Γ as well. The actual width is about 2500 MeV.

b) We begin with a formula for the differential cross-section,

$$\frac{d\sigma}{d\Omega} = \frac{|\mathcal{M}|^2}{64\pi^2 s} . \tag{17.37}$$

(This is a formula well worth memorizing!) The Feynman diagram for the process in Figure 8.1 looks like two decay vertices joined by a propagator. In the same way, the matrix element is the product of two decay matrix elements and a propagator:

$$|\mathcal{M}| = |\mathcal{M}|_{e\bar{e}} |\mathcal{M}|_{\nu\bar{\nu}} \left(\frac{1}{s - M_Z^2} \right) . \tag{17.38}$$

The decay matrix elements are in turn related to the widths of their respective channels by (17.33). Assuming that the angular integrations give 4π, this leads to

$$\sigma(s) = 16\pi \left(\frac{M_Z^2}{s} \right) \frac{\Gamma_{e\bar{e}} \Gamma_{\nu\bar{\nu}}}{(s - M_Z^2)^2} . \tag{17.39}$$

c) Displacing the pole leads directly to an expression for the total cross-section,

$$\sigma_{tot}(s) = 16\pi \left(\frac{M_Z^2}{s} \right) \frac{\Gamma_{e\bar{e}} \Gamma}{(s - M_Z^2)^2 + M_Z^2 \Gamma^2} . \tag{17.40}$$

On resonance, the cross-section becomes

$$\sigma_{tot}(s = M_Z^2) = 16\pi \frac{\Gamma_{e\bar{e}}\Gamma}{M_Z^2 \Gamma^2} \;. \tag{17.41}$$

Since $[(c_A^e)^2 + (c_V^e)^2] \approx 0.5[(c_A^\nu)^2 + (c_V^\nu)^2]$, $\Gamma_{e\bar{e}} \approx 0.5\Gamma_{\nu\bar{\nu}}$. Also, $\Gamma \approx 15\Gamma_{\nu\bar{\nu}}$, so

$$\sigma_{tot}(s = M_Z^2) \approx \frac{16\pi}{30} \frac{1}{M_Z^2}. \tag{17.42}$$

Inserting two powers of $\hbar c \approx 200$ MeV fm to get the right units yields a cross-section of

$$\sigma_{tot}(s = M_Z^2) \approx 8 \times 10^{-6} \text{ fm}^2 = 8 \times 10^{-2} \; \mu \text{ barns} \;. \tag{17.43}$$

d) One year is about 3×10^7 seconds, so the SLC should see about $\sigma L t \approx 10^6$ Z^0 events.

Solution 8.6.   a) The neutrinos are created via the process

$$e^- + p \to n + \nu_e \,, \tag{17.44}$$

and through the similar process,

$$e^+ + n \to p + \bar{\nu}_e \,. \tag{17.45}$$

They are detected through the reverse reactions:

$$\nu_e + n \to p + e^- \,,$$
$$\bar{\nu}_e + p \to n + e^+ \,. \tag{17.46}$$

Note that the neutrinos are *not* detected from $\nu_e e \to \nu_e e$ scattering in this experiment, since we are told that there are recoil nucleons.

b) We can write the velocity of the neutrinos v as a function of energy E and mass m, (letting $c = 1$):

$$v = \left(1 - \frac{1}{\gamma^2}\right)^{1/2} = \left(1 - \frac{m^2}{E^2}\right)^{1/2} \approx 1 - \frac{m^2}{2E^2}. \qquad (17.47)$$

The spread in arrival times in terms of the spread in arrival energies is therefore:

$$\Delta t = D\frac{m^2}{2}(E_1^{-2} - E_2^{-2}). \qquad (17.48)$$

We use the given data, namely $\Delta t \leq 1$ s, $D = 1.7 \times 10^5$ light years, $E_1 = 5$ MeV, and $E_2 = 20$ MeV to find a limit on the neutrino mass of $m_\nu \leq 3$ eV.

c) To estimate the total energy, we need to estimate the total number N of neutrinos emitted in the supernova. We start by calculating the integrated neutrino flux at the detector F_ν from the number of neutrinos observed:

$$10 = F_\nu \sigma N_t, \qquad (17.49)$$

where σ is the cross-section for the detection process and N_t is the number of target particles. To find an approximation for σ, recall that cross-sections for weak interactions at low energies are typically proportional to $G_F^2 s$, where s is the square of the center-of-mass energy. This leads to

$$\sigma \approx G_F^2 s \approx G_F^2 E_\nu^2 \approx 4 \times 10^{-46}\ m^2. \qquad (17.50)$$

We will assume the reactions are on the hydrogen nuclei in the water, since their recoil is more easily observed than the recoil of the oxygen nuclei. Then we have

$$N_t \approx (10^9\ g)\left(\frac{2}{18}\right)(mol/g)(N_A/mol) \approx 7 \times 10^{31}, \qquad (17.51)$$

where $N_A = 6 \times 10^{23}$ is Avogadro's number, 2 is the atomic mass of hydrogen, and 18 is the atomic mass of water. From equation (17.49), the integrated flux is

$$F_\nu = \frac{10}{\sigma N_t} \approx 3 \times 10^{14}\ m^{-2}. \qquad (17.52)$$

The detector is a distance D from the supernova. We assume that the flux was emitted isotropically, so that the total number of neutrinos emitted from the supernova is $N \approx F_\nu(4\pi D^2) \approx 10^{58}$. Since the average neutrino energy was 10 MeV, this implies

$$E_{tot} \approx 10^{59} \text{ MeV} \approx 10^{53} \text{ ergs} . \qquad (17.53)$$

d) The mass eigenstates are the states that propagate with a simple time dependence:

$$|\nu_1(t)\rangle = |\nu_1\rangle e^{-iE_1 t}, \quad |\nu_2(t)\rangle = |\nu_2\rangle e^{-iE_2 t}, \qquad (17.54)$$

where we have taken $\hbar = 1$. We can expand the wavefunctions for ν_a and ν_b as

$$
\begin{aligned}
|\nu_a\rangle &= \quad \cos\theta|\nu_1\rangle + \sin\theta|\nu_2\rangle , \\
|\nu_b\rangle &= -\quad \sin\theta|\nu_1\rangle + \cos\theta|\nu_2\rangle .
\end{aligned}
\qquad (17.55)
$$

Let the wavefunction of the neutrino which travels from the supernova to earth be $|\nu(t)\rangle$. The initial condition $|\nu(0)\rangle = |\nu_a\rangle$ gives

$$|\nu(t)\rangle = \cos\theta|\nu_1\rangle e^{-iE_1 t} + \sin\theta|\nu_2\rangle e^{-iE_2 t} . \qquad (17.56)$$

The probability P of finding ν_a at time t is

$$
\begin{aligned}
P(t) &= |\langle\nu_a|\nu(t)\rangle|^2 \\
&= \sin^4\theta + \cos^4\theta + 2\sin^2\theta\cos^2\theta\cos(\Delta E t) \\
&= (\sin^2\theta + \cos^2\theta)^2 - 4\sin^2\theta\cos^2\theta\sin^2(\Delta E t/2) \\
&= 1 - (\sin^2 2\theta)(\sin^2(\Delta E t/2)) ,
\end{aligned}
\qquad (17.57)
$$

with $\Delta E = E_2 - E_1$.

Before we go further, we would like to point out two assumptions that we make. As is conventional, we assume the neutrino wavefunction is a superposition of mass eigenstates with the same momentum p but different energies (see Bahcall). Secondly, since the different neutrino types have different masses (but the same momentum), they will travel with different velocities. If they start out in a localized wave packet, the mass eigenstates will eventually separate and no longer overlap. The distance over which this happens is known as the coherence length, and we will assume that this is much larger than the distance to the supernova.

We wish to write the energy difference ΔE in terms of the masses m_1 and m_2. Since

$$E = \sqrt{p^2 + m^2} \approx p + \frac{m^2}{2p}, \tag{17.58}$$

we can find

$$\Delta E \approx \frac{m_2{}^2 - m_1{}^2}{2p} \approx \frac{m_2{}^2 - m_1{}^2}{2E_\nu}, \tag{17.59}$$

where we have replaced $1/2p$ with $1/2E_\nu$, where E_ν is the average neutrino energy. This leads to a final form for the probability of finding ν_a at a distance d from the supernova,

$$P(\nu_a) \approx 1 - \sin^2(2\theta) \left[\sin^2 \left(\frac{m_2^2 - m_1^2}{4E_\nu} d \right) \right]. \tag{17.60}$$

The oscillation length is the length d_0 such that the argument of the second sine is equal to π,

$$\frac{m_2{}^2 - m_1{}^2}{4E_\nu} d_0 = \pi. \tag{17.61}$$

We assume that d_0 is greater than the distance d from the supernova to the earth so that we have

$$(m_2{}^2 - m_1{}^2) < \frac{\pi}{d} 4E_\nu. \tag{17.62}$$

We reinstate units to find a limit on the mass difference of

$$(m_2{}^2 - m_1{}^2) < 1.5 \times 10^{-20} \text{ eV}^2. \tag{17.63}$$

Solution 8.7.   a) The multiplicity factors arise from considerations of color and spin. The partial width $\Gamma_{u\bar{d}}$ is calculated by *summing* over all possible spin and color states of the decay products u and $\bar{d}$, and *averaging* over the W^+ spins. However, when these two quarks annihilate to form a W^+, we must average over the initial spin and color states, and sum over the W^+ spins. To convert the implicit sum over spin and color in $\Gamma_{u\bar{d}}$ to an average, we divide by $N_i = 2^2 \cdot 3^2 = 36$. To change the average over W^+ spins to a sum, we multiply $\Gamma_{u\bar{d}}$ by $N_W = 2J + 1 = 3$.

b) We can find the decay widths roughly using dimensional arguments. In Fermi theory, the width is proportional to the Fermi constant $G_F \approx 1.2 \times 10^{-5} \text{GeV}^{-2}$. The only energy scale in the problem is M_W (since at these energies the masses of the quarks and leptons can be neglected). So $\Gamma_{W \to e\nu} \approx G_F M_W^3 \approx 7$ GeV. (Note: A careful solution of this problem would find that we have missed a factor of $6\pi\sqrt{2}$—not of order unity! The more careful calculation gives $\Gamma_{W \to e\nu} = 232$ MeV. See Perkins, Appendix H.)

With the assumption (a good one) that the masses of the u and $\bar{d}$ quarks are negligible, we have that $\Gamma_{u\bar{d}} = 3\Gamma_{e\nu}$, where the factor of three comes from counting the color-anticolor states available. The total width for the decay includes contributions from three quark-lepton families. However, the W cannot have a top quark among its decay products because the top is too massive. Thus, with the assumption that the masses of all the other decay poducts can be neglected, we have $\Gamma_{tot} = 9\Gamma_{e\nu}$. This leads to an estimate for the W^+ lifetime of $\tau = 1/\Gamma_{tot} \approx 10^{-26}$ s, using $\Gamma_{e\nu} \approx 7$ GeV as found above. If instead we use the correct value $\Gamma_{e\nu} = 232$ MeV, we find $\tau \approx 3 \times 10^{-25}$ s. At the time of writing, experiments yield a value of τ which is within a factor of two of the predicted value.

c) In the center-of-mass frame, the proton and antiproton have the same momentum k (in opposite directions!). At high energies we have $E_p \approx k$, so we can write the proton four-momentum as $q = k(1, \hat{n})$ and the antiproton four-momentum as $q' = k(1, -\hat{n})$. So

$$s = (q + q')^2 = 4k^2. \tag{17.64}$$

In this frame we also have

$$\hat{s} = \left(\frac{x_p k + x_{\bar{p}} k}{x_p k - x_{\bar{p}} k} \right)^2 = (x_p + x_{\bar{p}})^2 k^2 - (x_p - x_{\bar{p}})^2 k^2 = x_p x_{\bar{p}} s. \quad (17.65)$$

d) We need to evaluate the expression for the total cross-section (8.5). First we evaluate the term $\sigma(\hat{s} = M_W^2)$ using the Breit-Wigner formula (8.4) and the results of part (a) and (b):

$$\sigma(\hat{s} = M_W^2) \approx \frac{\pi}{36 M_W^2}. \quad (17.66)$$

The total cross-section can be expressed as an integral over the quark momentum probabilities:

$$\sigma_{tot} \approx \frac{\pi \Gamma_{ev}}{3 M_W} \int_0^1 \int_0^1 dx_p \, dx_{\bar{p}} \, u(x_p) \bar{d}(x_{\bar{p}}) \delta(\hat{s} - M_W^2). \quad (17.67)$$

To do the integral over $x_{\bar{p}}$, we rewrite the delta function:

$$\delta(\hat{s} - M_W^2) = \delta(x_p x_{\bar{p}} s - M_W^2) = \frac{1}{x_p s} \delta\left(x_{\bar{p}} - \frac{M_W^2}{x_p s} \right). \quad (17.68)$$

Subsituting this into equation (17.67) and performing the integration over $x_{\bar{p}}$, we find

$$\sigma_{tot} = \frac{6\pi \Gamma_{ev}}{M_W s} \int_{M_W^2/s}^1 dx_p \, (1 - x_p)^2 (1 - \frac{M_W^2}{x_p s})^2 x_p^{-1}. \quad (17.69)$$

The integration limits come from the conditions that $x_p = M_W^2/x_{\bar{p}} s$ and $0 \le x_{\bar{p}} \le 1$. We could work out the integral exactly, or, in the spirit of our approximations, note that it is a dimensionless integral of order unity and drop it. The latter course gives us

$$\sigma_{tot} \approx \frac{6\pi \Gamma_{ev}}{M_W s} \approx 2 \times 10^{-33} \text{cm}^2 = 2 \times 10^{-3} \mu\text{barns}. \quad (17.70)$$

If L and ϵ are the collider luminosity and efficiency and t is the duration of the experiment, then the number of observed events is

$$\begin{aligned} N &= L\sigma \epsilon t = (10^{29} \text{cm}^{-2}\text{s}^{-1}) \times (2 \times 10^{-33} \text{cm}^2) \times 1 \times (1 \text{ year}) \\ &\approx 6000. \end{aligned} \quad (17.71)$$

Solution 8.8. a) Set $\hbar = c = 1$ in this problem. The decay of a B^+ meson occurs when the $\bar{b}$ quark decays weakly to a $\bar{u}$ or $\bar{c}$ quark as illustrated in Figure 17.5a. Note that since the mass of the top quark is

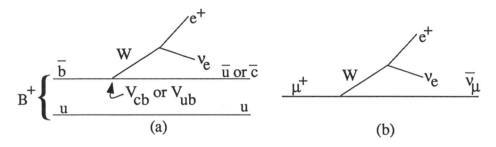

Figure 17.5.

greater than that of the bottom quark, the decay $\bar{b} \to \bar{t}$ is not allowed. The u quark of the B meson is a spectator quark and plays no part in the weak decay shown in Figure 17.5a, but subsequently the $\bar{u}$ and u or $\bar{c}$ and u will hadronize into a shower of particles. We assume this does not effect the decay shown. Thus the partial width of the decay $B^+ \to e^+\nu$+hadrons is approximately equal to $\Gamma(b \to ce\nu) + \Gamma(b \to ue\nu)$. However, we are told that $\Gamma(b \to u) \ll \Gamma(b \to c)$, so we can write

$$\Gamma(b \to ce\nu) \approx \Gamma(B^\pm \to e^\pm\nu + \text{hadrons}) \tag{17.72}$$

$$= \frac{1}{\tau_B}\text{BR}(B^\pm \to e^\pm\nu + \text{hadrons}) = 9 \times 10^{10} \text{ s}^{-1},$$

using information given in the question.

Now we need to extract a value for the matrix element $|V_{cb}|$. We can do this by comparing B decay to the very similar process of muon decay, which is shown in Figure 17.5b. The muon decay width is proportional to some vertex factor multiplied by the phase space, and the latter is proportional to m_μ^5. We can write

$$\Gamma(\mu^+ \to e^+\nu_e\bar{\nu}_\mu) = \frac{1}{\tau_\mu} = km_\mu^5, \tag{17.73}$$

for some constant k. Then, by analogy,

$$\Gamma(b \to e\nu c) = |V_{cb}|^2 km_b^5, \tag{17.74}$$

where k is the same constant as in equation (17.73), but now we take into account the extra factor of $|V_{cb}|^2$ arising from the mixing matrix.

We can rearrange equations (17.73) and (17.74) to find

$$|V_{cb}|^2 = \Gamma(b \to e\nu c)\tau_\mu \left(\frac{m_\mu}{m_b}\right)^5. \tag{17.75}$$

Then using the values given for m_b, m_μ, and τ_μ, we find $|V_{cb}| \approx 0.028$.

b) The question gives us

$$\frac{\Gamma(b \to u)}{\Gamma(b \to c)} < 0.08. \tag{17.76}$$

$\Gamma(b \to u)$ will be given by equation (17.74) if we replace $|V_{cb}|$ by $|V_{ub}|$. Thus we can find an upper limit: $|V_{ub}|^2 < 0.08|V_{cb}|^2$, or $|V_{ub}| < 0.008$.

c) The unitarity of the KM matrix requires that the sum of squares of any column or row equal one. We have found $|V_{cb}|$ and a limit on $|V_{ub}|$, so we calculate that $|V_{tb}| \geq 0.999$. We next count the number of ways the top can decay to $b + X$, where X is anything. To do this we need only the distinct states into which the W boson can decay. The total is (2 quarks)(3 colors) + (3 leptons) = 9, so the t quark has nine channels available. The muon, on the other hand, has only a single channel. In analogy with the arguments used in (a), the lifetime is

$$\tau_t = \frac{1}{9}\tau_\mu \left(\frac{m_\mu}{m_t}\right)^5 \frac{1}{|V_{tb}|^2} \approx 3 \times 10^{-20} \text{ seconds.} \tag{17.77}$$

Solution 8.9. The calculation of $\Gamma(H \to f\bar{f})$ is about the simplest possible real example of calculating a decay from a Feynman diagram. A good reference for the methods, conventions, and formulae used in this problem is *Quarks and Leptons* by Halzen and Martin.

As always, we set $\hbar = c = 1$. We need the matrix element for the decay $H \to f\bar{f}$ shown in Figure 17.6, where we have labeled the lines

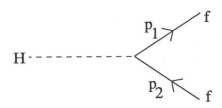

Figure 17.6.

with four-momenta p_1 and p_2. From the diagram, we can write down the matrix element as

$$\mathcal{M}_{ss'} = m_f(\sqrt{2}G_F)^{1/2}\sum_\alpha \bar{u}^s_\alpha(p_1)v^{s'}_\alpha(p_2), \qquad (17.78)$$

where we have used the vertex factor given in the question, and ignored possible factors of i which vanish when we take the modulus. The functions $u(p)$ and $v(p)$ are the usual four-component spinors associated with the plane-wave solutions of the Dirac equation for electrons and positrons respectively. The upper index on u and v is the spin index indicating whether the spinor describes a fermion of helicity $+1$ or -1, and the lower index labels the component of the spinor. If we are not measuring the spins of the decay fermions, then we must sum $|\mathcal{M}|^2$ over the final spins so that

$$|M|^2 \equiv \sum_{\substack{\text{all spins}}} |\mathcal{M}|^2 = \sqrt{2}G_F \sum_{ss'}\sum_{\alpha\beta} m_f^2 \bar{u}^s_\alpha(p_1)v^{s'}_\alpha(p_2)\bar{v}^{s'}_\beta(p_2)u^s_\beta(p_1) \cdot$$

$$(17.79)$$

This formula is often written in the compact notation

$$|M|^2 = \sqrt{2}G_F m_f^2 \text{Tr}\left[\bar{u}(p_1)v(p_2)\bar{v}(p_2)u(p_1)\right]. \qquad (17.80)$$

Using the completeness relations that $\sum_s u^s(p)\bar{u}^s(p) = \not{p} + m$ and $\sum_s v^s(p)\bar{v}^s(p) = \not{p} - m$, we find

$$|M|^2 = \sqrt{2}G_F m_f^2 \text{Tr}[(\not{p}_1 + m_f)(\not{p}_2 - m_f)], \qquad (17.81)$$

where the trace is now over only the spinor indices. Since $m_f \ll m_H$, the components of $\not{p}_1$ will be much greater than m_f and we can approximate $m_f \approx 0$. With this approximation,

$$|M|^2 = \sqrt{2}G_F m_f^2 \text{Tr}[\not{p}_1 \not{p}_2] = 4\sqrt{2}G_F m_f^2 p_1 \cdot p_2, \qquad (17.82)$$

where we have used the trace identity $\text{Tr}(\not{p_1} \not{p_2}) = 4p_1 \cdot p_2$. The four-momenta are

$$p_1 = \frac{m_H}{2}(1, \hat{n}), \quad p_2 = \frac{m_H}{2}(1, -\hat{n}), \tag{17.83}$$

where $\hat{n}$ is a unit vector. Taking the dot product of the momenta then gives

$$|M|^2 = 2\sqrt{2}G_F m_f^2 m_H^2. \tag{17.84}$$

To connect the matrix element with the decay rate, we use the formula for the decay from rest of a particle A to two particles labeled as 1 and 2:

$$\Gamma(A \to 1 + 2) = \frac{p_F}{32\pi^2 m_A^2} \int |M|^2 \, d\Omega. \tag{17.85}$$

The variable p_F is the magnitude of the final three-momentum of particle 1 or 2, and $d\Omega$ is an infinitesimal element of solid angle. Substitution of $m_A = m_H$, $p_F = m_H/2$, and $|M|^2$ from equation (17.84) yields

$$\Gamma_{f\bar{f}} \equiv \Gamma(H \to f\bar{f}) = \frac{\sqrt{2}G_F}{8\pi} m_f^2 m_H. \tag{17.86}$$

The width of the decay $\Gamma_{e^+e^-}$ is then 7×10^{-12} MeV (where we have used the values $G_F \approx 10^{-5}$ GeV^{-2} and $m_e = 0.5$ MeV). The lifetime τ from the decay to e^+e^- is therefore

$$\tau = \hbar/\Gamma(H \to e^+e^-) \approx 10^{-10}\text{s}. \tag{17.87}$$

b) According to the relativistic Breit-Wigner formula, the cross-section near resonance for the process $e^+e^- \to H \to f\bar{f}$ is given by

$$\sigma = \left(\frac{4\pi s}{k^2}\right)\left(\frac{2J+1}{(2S_a+1)(2S_b+1)}\right)\frac{\Gamma_{e^+e^-}\Gamma_{f\bar{f}}}{(s-m_H^2)^2 + m_H^2\Gamma^2}, \tag{17.88}$$

where Γ is the total width, S_a and S_b are the spins of the incoming particles, J is the spin of the resonance, s is the square of the center-of-mass energy, and k is the magnitude of the fermion three-momentum. Setting $S_a = S_b = 1/2$, $J = 0$, and, since we are at resonance, $s = 4k^2 = m_H^2$, we find

$$\sigma = \left(\frac{4\pi}{m_H^2}\right)\frac{\Gamma_{e^+e^-}\Gamma_{f\bar{f}}}{\Gamma^2}. \tag{17.89}$$

Since the width is dominated by decay to a $b\bar{b}$ pair, $\Gamma \approx 3\Gamma_{b\bar{b}}$, where the factor of 3 comes from the fact that the $b\bar{b}$ can take any one of three color-anticolor combinations. The cross-section for this reaction is therefore

$$\sigma = \frac{4\pi}{m_H^2}\left(\frac{\Gamma_{e^+e^-}}{\Gamma_{b\bar{b}}}\right) = \frac{4\pi}{m_H^2}\left(\frac{m_e^2}{3m_b^2}\right) = 1.7 \times 10^{-17}\ \text{MeV}^{-2}, \qquad (17.90)$$

where we have taken $m_b = 5$ GeV and $m_H = 50$ GeV. Reinstating c and $\hbar$, we find $\sigma = (1.7 \times 10^{-17}\ \text{MeV}^{-2})(197\ \text{MeV fm})^2 = 7 \times 10^{-39}\text{cm}^2$. The number of events in one year is $\sigma \times (1\ \text{year}) \times (\text{luminosity}) \approx 4$ events.

Solution 8.10.   a) We can think of the charged-current interaction in terms of the incoming neutrino emitting a W^+ boson and changing into a negatively charged muon. The W^+ can only be absorbed when the target quark is a d, which is converted into a u, and so there is just one Feynman diagram for this process (Figure 17.7a).

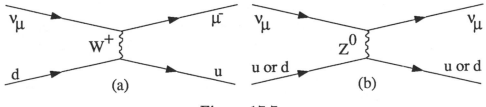

Figure 17.7.

The neutral-current interaction takes place via a Z^0 boson, and both the incoming and target fermions retain their flavors. The neutrino can scatter off both u and d quarks, and thus there are two equivalent diagrams contributing to the process (Figure 17.7b).

b) Our first step is to integrate the expression given for $d\sigma/dy$ over the range of y, namely $0 \leq y \leq 1$, to get the total cross-section:

$$\sigma_{tot} = \begin{cases} \dfrac{G_F^2\, s}{\pi} \overline{Q_W^2}\,, & \text{left-handed,} \\[4mm] \dfrac{G_F^2\, s}{3\pi} \overline{Q_W^2}\,, & \text{right-handed.} \end{cases} \qquad (17.91)$$

Now consider charged-current scattering. We will assume that the target is unpolarized, which means that it contains equal numbers of left- and right-handed quarks. Furthermore, since the target is composed of equal numbers of protons and neutrons, there will be equally many u and d quarks. Since in this experiment charged-current scattering can only take place off d quarks, when we average the square of the weak charge over u and d, we get $\overline{L_{CC}^2} = 1/2$, and therefore the total charged-current cross-section, averaged over quark polarizations, is

$$\sigma_{CC} = \frac{G_F^2 s}{\pi}\frac{1}{2}\left(\overline{L_{CC}^2} + \frac{1}{3}\overline{R_{CC}^2}\right) = \frac{G_F^2 s}{4\pi}\,. \qquad (17.92)$$

In the case of neutral-current scattering, we have to be very careful to assign the correct values of weak isospin to the particles. In particular, u and d have $I_3 = +1/2, -1/2$ respectively for the left-handed components of the spinor field, and $I_3 = 0$ for the right-handed components of both. Thus although L_{NC} and R_{NC} are given by the same formula, the actual values of the weak charges are very different. This is one of the fundamental features of the standard model, and is the source of parity violation in weak interactions. Note also that Q_e is $+2/3$ for u and $-1/3$ for d.

Therefore, for the left-handed fermions, u has weak charge $L_{NC} = 1/2 - 2/3 \sin^2 \theta_W$, and d has $L_{NC} = -1/2 + 1/3 \sin^2 \theta_W$. The right-handed u has $R_{NC} = -2/3 \sin^2 \theta_W$, and the d has $R_{NC} = 1/3 \sin^2 \theta_W$. Therefore the left- and right-handed weak charges, averaged over target quarks, are

$$\overline{L_{NC}^2} = \frac{1}{2}\left[\left(\frac{1}{2} - \frac{2}{3}\sin^2\theta_W\right)^2 + \left(-\frac{1}{2} + \frac{1}{3}\sin^2\theta_W\right)^2\right],$$

$$\overline{R_{NC}^2} = \frac{1}{2}\left[\left(\frac{2}{3}\sin^2\theta_W\right)^2 + \left(\frac{1}{3}\sin^2\theta_W\right)^2\right]. \qquad (17.93)$$

The total neutral-current cross-section is the average of the left- and right-handed cross-sections,

$$\sigma_{NC} = \frac{1}{2}\frac{G_F^2 s}{\pi}\left(\overline{L_{NC}^2} + \frac{1}{3}\overline{R_{NC}^2}\right) = \frac{G_F^2 s}{4\pi}\left(\frac{1}{2} - \sin^2\theta_W + \frac{20}{27}\sin^4\theta_W\right).$$

$$(17.94)$$

To find $\sin^2\theta_W$ we take the ratio of the cross-sections (17.92) and (17.94), and equate it with the given value,

$$R_\nu = 0.3 = \frac{1}{2} - \sin^2\theta_W + \frac{20}{27}\sin^4\theta_W.$$ (17.95)

This is just a quadratic equation which we can solve (discarding the unphysical root which is greater than unity) to find

$$\sin^2\theta_W = 0.24.$$ (17.96)

This is a very good estimate of the current experimental value, which is around 0.23.

Chapter 18

Atomic & General Physics—Solutions

Solution 9.1. a) The Bohr theory of the hydrogen atom can be directly applied to positronium after noting that the reduced mass for positronium has the value $\mu = m_e/2$, rather than $\mu \approx m_e$ as for the hydrogen atom. If we ignore fine structure, the energy levels for positronium are given by

$$E_n = -\frac{1}{2}\alpha^2\mu c^2 \left(\frac{1}{n^2}\right) \equiv -\frac{1}{2n^2}\, \text{Ry}, \tag{18.1}$$

where $\alpha \approx 1/137$ is the fine-structure constant, and a Rydberg is an energy unit with 1 Ry= 13.6 eV, the ground state energy of hydrogen. Thus, the ground state ($n = 1$) binding energy for positronium is half that of hydrogen, or 6.8 eV. From equation (18.1), the $2p \to 1s$ transition corresponds to an energy difference of 5.1 eV, or a wavelength of 2400 Å.

b) The decay rate Γ for electric dipole transitions between two states is proportional to $\omega^3 |\langle \mathbf{r}\rangle|^2$, where $\hbar\omega$ is the energy difference between the states, $\mathbf{r}$ is the relative coordinate between the positron and the electron, and the constant of proportionality depends only on natural constants and pure numbers. From part (a), we see that ω is half as

283

large for a given transition in positronium as in hydrogen. The effective Bohr radius in positronium, $a_p \equiv \hbar^2/\mu e^2$, is equal to twice the Bohr radius in hydrogen. Since all lengths in the atom scale with the effective Bohr radius, the matrix element of $\mathbf{r}$ between the $1s$ and $2p$ states of positronium will be twice the matrix element of $\mathbf{r}$ between the same states in hydrogen. Combining these factors of two, we find that the lifetime, or $1/\Gamma$, will be twice as large in positronium as in hydrogen, or 3.2 ns.

c) and d) The magnetic field on the electron due to the positron's magnetic moment is analogous to the magnetic field on the electron due to the proton in hydrogen . Therefore, part (d) is asking for an estimate of the "hyperfine" splitting in positronium. A naive estimate of the dipole magnetic field on the electron would be μ_B/a_p^3, and the corresponding energy splitting due to this field would be on the order of $\mu_B(\mu_B/a_p^3)$. On dimensional grounds these must be the correct answers, aside from numerical factors. Similarly, the splitting in the $1s$ state of hydrogen must look like $\mu_B(\mu_P/a_H^3)$ with the same number out front. The magnetic moment of the proton is $\mu_P = g_P\mu_N/2$, where $g_P \approx 5.6$, and μ_N is the nuclear magneton (just as the e^- has magnetic moment $g_s\mu_B/2$ with $g_s \approx 2$). Recalling that the hyperfine splitting in the ground state of hydrogen gives rise to the 21 cm line, important in astronomy, we can estimate the hyperfine splitting in positronium as

$$\Delta E \approx \frac{2\pi\hbar c}{(21 \text{ cm})}\left(\frac{\mu_B}{\mu_P}\right)\left(\frac{a_H^3}{a_p^3}\right) = \frac{2\pi\hbar c}{(21 \text{ cm})}\left(\frac{g_s m_P}{g_P m_e}\right)\left(\frac{1}{8}\right) = 5 \times 10^{-4} \text{ eV},$$
(18.2)

where $m_P \approx 1$ GeV is the mass of the proton. While this estimate is a good one, the actual cause of the hyperfine splitting in the $1s$ state is somewhat subtle and is outlined below.

According to the general prescription for finding hamiltonians, we write down the classical hamiltonian for the interacting magnetic dipoles of the e^- and e^+ and then interpret the separation $\mathbf{r}$ and the magnetic moments as quantum-mechanical operators. Classically the magnetic field at a position $\mathbf{r}$ from the positron's dipole moment $\boldsymbol{\mu}_{e+}$ is

$$\mathbf{B}(\mathbf{r}) = \frac{1}{r^3}\left(\frac{3(\mathbf{r}\cdot\boldsymbol{\mu}_{e+})\mathbf{r}}{r^2} - \boldsymbol{\mu}_{e+}\right) + \frac{8\pi}{3}\boldsymbol{\mu}_{e+}\delta^3(\mathbf{r}).$$
(18.3)

The interaction energy with the electron magnetic dipole $\boldsymbol{\mu}_{e-}$ is $[-\boldsymbol{\mu}_{e-} \cdot \mathbf{B(r)}]$. With the substitutions

$$\boldsymbol{\mu}_{e+} = +\frac{g_s\mu_B}{\hbar}\mathbf{S}_1 \quad \text{and} \quad \boldsymbol{\mu}_{e-} = -\frac{g_s\mu_B}{\hbar}\mathbf{S}_2, \tag{18.4}$$

where $\mathbf{S}_1$ and $\mathbf{S}_2$ are the spin of the positron and electron, respectively, and $\mathbf{r}$ is reinterpreted as a vector operator, we have the following hyperfine hamiltonian:

$$H = \frac{g_s^2\mu_B^2}{\hbar^2 r^3}\left[\frac{3(\mathbf{r}\cdot\mathbf{S}_1)(\mathbf{r}\cdot\mathbf{S}_2)}{r^2} - \mathbf{S}_1\cdot\mathbf{S}_2\right] + \frac{8\pi}{3}\left(\frac{g_s\mu_B}{\hbar}\right)^2|\psi(0)|^2\,\mathbf{S}_1\cdot\mathbf{S}_2, \tag{18.5}$$

where $|\psi(0)|^2$ is the probability that the electron and positron are at the same position. Since the $1s$ state is spherically symmetric, the expectation value of the term in brackets is easily shown to vanish for both the singlet and triplet states. The second term in the hamiltonian is called the "Fermi contact interaction," and does not vanish for s-states. We rewrite the Fermi contact interaction as

$$H_f = \frac{8\pi}{6}\left(\frac{g_s\mu_B}{\hbar}\right)^2|\psi(0)|^2\,(\mathbf{S}^2 - \mathbf{S}_1^2 - \mathbf{S}_2^2) \tag{18.6}$$

$$= \frac{4\pi}{3}g_s^2\mu_B^2\,|\psi(0)|^2\,S(S+1) + \text{constant}, \tag{18.7}$$

where $\mathbf{S} = \mathbf{S}_1 + \mathbf{S}_2$, and by definition, $S = 0$ for the singlet state and $S = 1$ for triplet state. The energy difference between the singlet and triplet state is given by

$$\Delta E = \langle S = 1\,|H_f|\,S = 1\rangle - \langle S = 0\,|H_f|\,S = 0\rangle. \tag{18.8}$$

The splitting from the magnetic interaction is therefore

$$\Delta E = \frac{8\pi}{3}\left(\frac{1}{\pi a_p^3}\right)g_s^2\mu_B^2 = \frac{4\mu_B^2}{3a_H^3} = \frac{2}{3}\alpha^2\,\text{Ry} = 5\times10^{-4}\text{ eV}, \tag{18.9}$$

where we use the result $|\psi(0)|^2 = 1/\pi a_p^3$. This is just what we found by dimensional arguments in equation (18.2). The magnetic field experienced by the electron is therefore $\Delta E/\mu_B \approx 10^5$ gauss (recalling that $\mu_B/h = 1.4$ MHz/gauss).

There is a further contribution to the splitting in positronium due to an "annihilation diagram," with the value of $\alpha^2 \text{Ry}/2$, which we shall not derive (see Sakurai). This annihilation term is of course not present in hydrogen.

e) The question of correlation of photon polarizations forms the basis for the Einstein-Podolsky-Rosen paradox. There is a good discussion of the answer to this part of the question, the EPR paradox, and its resolution, in Sakurai, Chapter 4.

The parity of the singlet ^1S state is $-1(-1)^{L+S} = -1$, where the first factor of -1 is due to the opposite intrinsic parities of the electron and positron. Since parity is conserved by electromagnetic interactions, the two decay photons must also be in a state of negative parity. The total angular momentum of the $n = 1$ singlet state is zero, so the total angular momentum of the photons must be zero. The photons are therefore in a symmetric state with zero angular momentum and negative parity. Any correlation function must have these same properties, i.e., it must be a pseudoscalar. The only quantities available from which we can form the correlation function are the polarization vectors of each photon, and the relative momentum $\mathbf{k}$. The only way to combine three vectors into a pseudoscalar is to form a scalar triple product: $(\epsilon_1 \times \epsilon_2) \cdot \mathbf{k}$, or permutations. This has a maximum when the polarizations are at a relative angle of 90 degrees.

Solution 9.2. a) The lens serves two purposes. It defines an aperture which diffracts the incoming plane wave, and it brings the far-field diffraction pattern of this aperture into the focal plane (see Hecht and Zajac, Chapter 10). The far-field pattern can be found by considering the interference between infinitesimal area elements of the aperture. The resulting intensity distribution (for a circular aperture) is called an Airy pattern. This pattern has a main lobe centered on the optical axis, and a first null at a radius of

$$r = 1.22 \frac{f\lambda}{D}, \tag{18.10}$$

from the optical axis (which is a fact well worth memorizing). The radius of this "spot" is the basis for the Rayleigh criterion, which is a measure of the diffraction-limited resolution of any optical system, e.g., a telescope or a microscope.

A quick way to calculate the *approximate* size of the central spot is to consider diffraction by a square aperture. The top half of the square must interfere destructively with the bottom half at the position of the first null. A little ray tracing yields the result given above, without the factor of 1.22.

b) Let the slits lie in the xy-plane, parallel to the x-axis. Let them each have a length l, and the separation between them be a distance d. Consider first the diffraction in the x-direction. In that dimension, the light will focus down to a spot about the size of the Airy spot described in part (a), but with the slit length replacing the diameter of the lens. This gives a spot radius of about $\lambda f/l$. Consider now the diffraction in the y-direction. If we ignore the width of the slits, we can use the geometry shown in Figure 18.1 to find the condition for constructive interference, $d\sin\theta = n\lambda$. In the limit of small θ, this gives $\theta \approx n\lambda/d$.

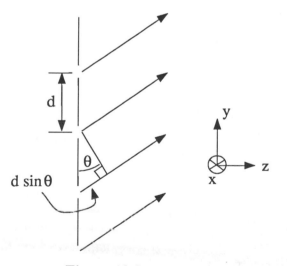

Figure 18.1.

At these angles, the constructive interference will cause bright spots to occur. The spacing ΔY between spots in the focal plane of the lens is $\Delta Y \approx n\lambda f/d$.

c) By tracing the two rays shown in Figure 18.2 (one passing un-deflected through the center of the lens, the other passing through its focal point), we can see from similar triangles that

$$\frac{x}{g} = \frac{y}{h} \quad \text{and} \quad \frac{x}{f} = \frac{y}{h-f},\qquad (18.11)$$

which leads to

$$\frac{1}{f} = \frac{1}{h} + \frac{1}{g},\qquad (18.12)$$

which is the well known lensmaker's formula.

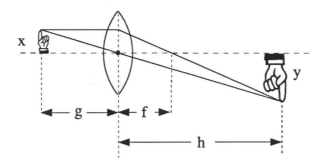

Figure 18.2.

d) When looking at an image, the human eye detects differences in the intensity (or the log of the intensity) of the electric field, not phase differences. The transparent microbe, unfortunately, only phase-modulates the field, rather than amplitude-modulating it. This section of the problem describes the useful technique of using a phase plate to change phase-modulated waves into amplitude-modulated waves, called phase contrast microscopy.

This part of the problem can best be understood by using an impor-tant result from diffraction theory, which is that the far-field diffraction pattern is the Fourier transform of the aperture function (see Hecht and Zajac, Chapter 11). The aperture function describes the ampli-tude and phase of the wave which is diffracting through the aperture, as a function of position in the aperture plane. As mentioned previ-ously, the far-field pattern appears at the focal plane of the lens (at

B). Different spatial frequencies present in the complex-valued aperture function show up as different bright spots at B; the higher the spatial frequency, the farther the spot lies from the optical axis. A uniformly illuminated aperture thus mainly has a bright spot on the optical axis (as in part (a)), while an aperture with a lot of small structure will have a lot of light on the focal plane away from the optical axis, as well as a bright spot on axis. The idea behind using the phase plate is that the microbe is small, and much of the information about its image will lie away from the optical axis at B, missing the phase plate. If the phase plate is about the size of the Airy spot, most of the light from the microbe containing information about its structure will not be phase-shifted by the phase plate, while the light corresponding to its uniform background will be phase-shifted. Therefore x should be about $\lambda f / D$.

To see how the phase plate enhances the image, consider a wave with an electric field $E_0 \hat{e} = (E_1 + E_2)\hat{e}$ impinging on the microbe, where E_0, E_1, and E_2 are all positive. Some fraction (E_1/E_0) of that wave is slightly phase-shifted by the microbe, but passes through the phase plate at the transform plane B. The rest, (E_2/E_0), is diffracted so that it misses the phase plate on the transform plane.

At the transform plane, the wave which passes through the phase plate is delayed by $\pi/2$, so that

$$E_1 e^{-i\omega t + \phi}\hat{e} \longrightarrow -i E_1 e^{-i\omega t + \phi}\hat{e} . \tag{18.13}$$

The part of the wave which is diffracted away from the phase plate is

$$E_2 e^{-i\omega t + \phi}\hat{e} . \tag{18.14}$$

If we assume the phase shift ϕ induced by the microbe is small, we can substitute $e^{i\phi} \approx (1 + i\phi)$ in both of the above expressions. Doing so, and combining the two fields at the image plane gives the electric field there,

$$\mathbf{E}_{image} = ((\phi E_1 + E_2) + i(\phi E_2 - E_1)) e^{-i\omega t}\hat{e} . \tag{18.15}$$

Squaring to find the intensity (and remembering to take the time average) gives

$$E_{image}^2 = \frac{1}{2}(1 + \phi^2)(E_1^2 + E_2^2) , \tag{18.16}$$

which is an amplitude-modulated, ϕ-dependent intensity.

If the center of the phase plate is opaque, the uniform background will be almost completely filtered out, and the image of the microbe will be light on a dark background.

Solution 9.3. a) Let r be the radial distance from the axis of symmetry of the lens and x be the thickness parallel to the axis of symmetry measured from the flat side of the lens (Figure 18.3). Consider a source

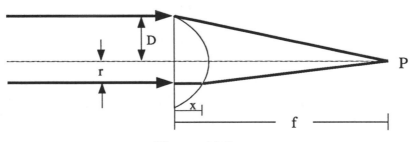

Figure 18.3.

placed a large distance to the left of the lens, so that rays incident on the lens from the source can be treated as parallel. Then the focusing condition is that the optical path length of all such rays from the source to the focal point must be the same. Two rays are drawn in Figure 18.3. The upper ray just grazes the top of the lens, while the lower ray exits the lens at the point (r, x). Setting the path lengths of the two rays equal gives

$$xn + \sqrt{r^2 + (f - x)^2} = \sqrt{f^2 + D^2/4}. \qquad (18.17)$$

Rearranging to find r as a function of x, we have

$$r^2 = \left(\sqrt{f^2 + D^2/4} - nx\right)^2 - (f - x)^2. \qquad (18.18)$$

It is straightforward (if tedious) to invert this to give x as a function of r, if desired.

b) We use the common trick for lens problems of finding the intersection of two rays with simple trajectories in order to find the focus
(see Figure 18.4). The first travels straight through the center of the

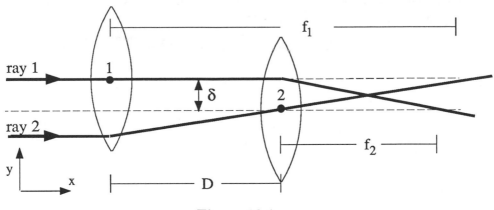

Figure 18.4.

first lens and is deflected by the second lens through the focal point f_2.
Using coordinates with the origin at the center of the second lens, this
path (after the second lens) can be written

$$y = \delta - \left(\frac{\delta}{f_2}\right) x \ . \tag{18.19}$$

The second ray is chosen to pass through the center of the second lens
and the focal point of the first, f_1. Its path after the second lens is

$$y = \left(\frac{\delta}{f_1 - D}\right) x \ . \tag{18.20}$$

The intersection of these two lines determines the focal point of the two
lens system. Some algebraic manipulation shows that the focal point
occurs at

$$x = \frac{f_2(f_1 - D)}{f_1 + f_2 - D}, \tag{18.21}$$

$$y = \frac{f_2\delta}{f_1 + f_2 - D}. \tag{18.22}$$

Solution 9.4.   Note that the fine structure constant is $\alpha \equiv e^2/\hbar c$ and the Bohr radius is $a_0 \equiv \hbar^2/m_e e^2$. In spite of the wording of the question, we will include factors of two where they are obvious.

a) The energy of the nth level is $E_n = -(\alpha^2 m_e c^2)/(2n^2)$. The transition of interest between the $n = 1$ and $n = 2$ levels has the energy $\Delta E = h f_a = \frac{3}{8} m_e c^2 \alpha^2$, so that the frequency of the absorption line f_a is

$$f_a = \frac{3}{16\pi} \frac{m_e c^2}{\hbar} \alpha^2 . \tag{18.23}$$

b) The expression for the linewidth due to spontaneous emission of electric dipole radiation is, in units of energy,

$$\Delta E = h\Delta f = \frac{4}{3} \left(\frac{\omega}{c}\right)^3 |er|^2 , \tag{18.24}$$

where $|er|$ is the dipole operator matrix element between the initial and final states, and ω is the angular frequency of the transition. We can approximate $|er|^2$ by $e^2 a_0^2$. Thus, with the expression for f_a from (18.23), we find that the linewidth is

$$\Delta f \approx \frac{9}{256\pi} \frac{m_e c^2}{\hbar} \alpha^5 . \tag{18.25}$$

c) The Doppler width is due to the thermal motion of the hydrogen atoms at temperature T. We may use the equipartition theorem to estimate their velocity: $m_H v^2/2 \approx 3kT/2$, so that $v \approx \sqrt{3kT/m_H}$, where m_H is the mass of hydrogen. In order not to be ionized, the cloud must be at a reasonably low temperature, so we can use the non-relativistic Doppler formula to find the frequency shift: $f_0/f \approx 1 \pm v/c$. Thus we may approximate the width of the observed frequency by $\Delta f_o \approx vf/c$, to within a numerical factor of order unity. We use the expression for f_a from equation (18.23), and find

$$\Delta f_o \approx \frac{3}{16\pi} \sqrt{\frac{3kT}{m_H}} \frac{m_e c}{\hbar} \alpha^2 . \tag{18.26}$$

d) From the uncertainty principle $\Delta E \Delta t \geq \hbar$, we estimate that the collisional linewidth is roughly $\Delta f_c \approx 1/\tau$, where τ is the mean time between collisions for one atom. We may write $\tau \approx l/\langle v \rangle$, with $\langle v \rangle$ the average speed and l the mean free path. We know from statistical mechanics that $l \approx 1/\rho \pi d^2$ for spherical objects of diameter d and number density ρ. We approximate d by $2a_0$. For $\langle v \rangle$, we again use the velocity due to thermal motion, so that

$$\Delta f_c \approx \pi \rho \sqrt{\frac{3kT}{m_H}} (2a_0)^2 \approx 4\pi \rho \sqrt{\frac{3kT}{m_H}} \left(\frac{\hbar c}{m_e c^2} \right)^2 \frac{1}{\alpha^2}. \qquad (18.27)$$

e) The fine structure splitting between states differing only in their total angular momentum j is proportional to $\alpha^2 E_n$. (Hence the name.) Thus, the frequency splitting between the $2p$ states with total angular momentum $j = 1/2$ and $j = 3/2$ is approximately

$$\Delta f_f \approx \alpha^2 \left(\frac{m_e c^2 \alpha^2}{h} \right) = \alpha^4 \frac{m_e c^2}{h}. \qquad (18.28)$$

The origin of this splitting is the spin-orbit term arising from the interaction of the magnetic moment of the electron with the magnetic field the electron sees as it moves through the electric field of the proton. A heuristic derivation of this term is given in Problem 9.7c.

f) The hyperfine splitting can be thought of as the interaction of the magnetic moment of the proton and the magnetic moment of the electron. The magnetic moment of the electron is $g_s \mu_B/2$ and of the proton is $g_P \mu_N/2$, where $g_s \approx 2$ and $g_P \approx 5.6$. The Bohr magneton is $\mu_B = e\hbar/2m_e c$ and the nuclear magneton is $\mu_N = e\hbar/2m_p c$. Since the electron and the proton are a distance a_0 apart, we estimate the interaction energy of their magnetic dipoles as

$$\Delta f_h \approx \frac{g_s g_P \mu_B \mu_N}{4a_0^3} \approx \alpha^4 \frac{m_e}{m_p} \frac{m_e c^2}{h}. \qquad (18.29)$$

The above argument amounts to nothing more than dimensional analysis. For a fuller discussion of the hyperfine interaction see Problems 9.1c and 9.5.

Solution 9.5.   To find the time evolution of the excited states, we need to know the energy eigenstates of the hamiltonian. The excited state $\psi(0)$ given in the question is not an eigenstate of the hyperfine hamiltonian. The eigenstates are labeled by the quantum number F of the total angular momentum $\mathbf{F} = \mathbf{J} + \mathbf{I}$, and the z-component of $\mathbf{F}$, denoted m_F. To show this we simply rewrite $\mathbf{I} \cdot \mathbf{J}$ as $(\mathbf{F}^2 - \mathbf{I}^2 - \mathbf{J}^2)/2$. Then, using the values $I = 1/2$ and $J = 1/2$, we have

$$H_{hyp} = \frac{A\hbar}{2}\left(F(F+1) - \frac{6}{4}\right). \tag{18.30}$$

Clearly H_{hyp} is diagonal in the basis $|I\ J\ F\ m_F\rangle$, which we abbreviate as $|F\ m_F\rangle$. Therefore the hyperfine hamiltonian splits the otherwise fourfold degenerate excited state into a triplet $(F = 1)$ and a singlet $(F = 0)$ state, and splits the ground state similarly. Let us denote the excited states by $|F\ m_F\rangle_e$ and the ground states by $|F\ m_F\rangle_g$. The action of the full hamiltonian on these states is

$$H\,|F\ m_F\rangle_e = [E' + \tfrac{A\hbar F(F+1)}{2}]\,|F\ m_F\rangle_e \equiv E_F\,|F\ m_F\rangle_e\ , \tag{18.31}$$

$$H\,|F\ m_F\rangle_g = [E^0 + \tfrac{A\hbar F(F+1)}{2}]\,|F\ m_F\rangle_g\ , \tag{18.32}$$

where E^0 and E' are the parts of the energy that are independent of F, and E_F is defined as shown. For convenience, let us arbitrarily set E^0 equal to zero.

The initial wavefunction $\psi(0)$ is written in the basis $|J\ m_J\rangle\,|I\ m_I\rangle$, abbreviated here as $|m_J\ m_I\rangle$. Expanding the $|m_J\ m_I\rangle$ basis in terms of our eigenstates $|F\ m_F\rangle$ gives

$$|\tfrac{1}{2}\ \tfrac{1}{2}\rangle = |1\ 1\rangle\,, \qquad |\tfrac{1}{2}\ -\tfrac{1}{2}\rangle = \tfrac{1}{\sqrt{2}}\{|1\ 0\rangle + |0\ 0\rangle\}\,,$$
$$|-\tfrac{1}{2}\ \tfrac{1}{2}\rangle = \tfrac{1}{\sqrt{2}}\{|1\ 0\rangle - |0\ 0\rangle\}\,, \quad |-\tfrac{1}{2}\ -\tfrac{1}{2}\rangle = |1\ -1\rangle\,.$$

$$\tag{18.33}$$

Therefore, expanding $\psi(0)$ in terms of the energy eigenstates gives

$$\psi(t) = \sum_{F,m_F} C_{F,m_F}\,|F\ m_F\rangle\,\exp(-iE_Ft/\hbar)\exp(-t/2\tau)\,, \tag{18.34}$$

where $C_{11} = 1/\sqrt{2}$, $C_{10} = 1/2$, $C_{1-1} = 0$, $C_{00} = 1/2$. In the above expansion we have included the phenomenological factor $\exp(-t/2\tau)$

which gives an exponential decay of the excited states with a time constant τ.

Now we can address the question of the time dependence of the fluorescence. From Fermi's golden rule, we know that the decay rate is proportional to $|\langle\psi(t)|\epsilon\cdot\mathbf{D}|\psi_g\rangle|^2$, where ϵ is the polarization of the emitted photon and $\mathbf{D}$ is the dipole operator. A knowledge of the selection rules for electric dipole radiation will allow us to avoid unnecessary calculation of matrix elements that vanish. The selection rules are: $\Delta F = 0, \pm 1$, but not $F = 0 \rightarrow F = 0$, and $\Delta m_F = 0, \pm 1$. Since the excited beam caused transitions which increased m_J by one, decays of the excited state which decrease m_F by one will give light with polarization of the same handedness as the exciting laser beam. In other words, if $\hat{\epsilon}_+$ is the polarization vector for the incident beam, then we need to consider the operator $\mathbf{D}\cdot\hat{\epsilon}_+$, which will only connect states where $\Delta m_F = -1$.

Because we only observe the polarization of the photon, and not its energy, we cannot distinguish, for example, the decay of the $|1\ 1\rangle_e$ state to the $|1\ 0\rangle_g$ state from that of the $|1\ 1\rangle_e$ state to the $|0\ 0\rangle_g$ state. Therefore, we must add the matrix elements for all possible decays which yield $\Delta m_F = -1$, and then square the result to find the rate of emission of photons with the same handedness. From the sum in (18.34) and using only the final states consistent with the selection rules, we have, for the matrix element appearing in Fermi's golden rule,

$$
\begin{aligned}
M \;=\; & e^{-t/2\tau}\Big[C_{11}\langle 1\ 1\,|_e\,\mathbf{D}\cdot\hat{\epsilon}_+|\,1\ 0\rangle_g\, e^{i(E'+A\hbar)t/\hbar}e^{-iAt} \quad (18.35)\\
& + C_{11}\langle 1\ 1\,|_e\,\mathbf{D}\cdot\hat{\epsilon}_+|\,0\ 0\rangle_g\, e^{i(E'+A\hbar)t/\hbar}\\
& + C_{10}\langle 1\ 0\,|_e\,\mathbf{D}\cdot\hat{\epsilon}_+|\,1\ -1\rangle_g\, e^{i(E'+A\hbar)t/\hbar}e^{-iAt}\\
& + C_{00}\langle 0\ 0\,|_e\,\mathbf{D}\cdot\hat{\epsilon}_+|\,1\ -1\rangle_g\, e^{iE't/\hbar}e^{-iAt}\Big]\,.
\end{aligned}
$$

Here the energies of the states have been calculated from equations (18.31) and (18.32).

The operator $\mathbf{D}\cdot\hat{\epsilon}_+$ only acts on the electronic angular momentum J, and not on the nuclear spin. If we define

$$
\langle J = \tfrac{1}{2}\ m_J = \tfrac{1}{2}\,|_e\,\mathbf{D}\cdot\hat{\epsilon}_+|\,J = \tfrac{1}{2}\ m_J = -\tfrac{1}{2}\rangle_g \equiv \sqrt{2}k\,, \qquad (18.36)
$$

then using the inverse of the set of basis transformation given in equation (18.33), we can evaluate the individual matrix elements appearing in (18.35) to find

$$
\begin{aligned}
\langle 1\,1\,|_e\, \mathbf{D}\cdot\hat{\epsilon}_+|1\,0\rangle_g &= k\,, \\
\langle 1\,1\,|_e\, \mathbf{D}\cdot\hat{\epsilon}_+|0\,0\rangle_g &= -k\,, \\
\langle 1\,0\,|_e\, \mathbf{D}\cdot\hat{\epsilon}_+|1\,-1\rangle_g &= k\,, \\
\langle 0\,0\,|_e\, \mathbf{D}\cdot\hat{\epsilon}_+|1\,-1\rangle_g &= k\,.
\end{aligned}
\tag{18.37}
$$

Substituting these values into the expression for M (18.35) and squaring the result gives

$$
|M|^2 \propto \left[3 + \sqrt{2} - \cos At - \sqrt{2}\cos 2At\right] e^{-t/\tau}\,.
\tag{18.38}
$$

It is important to realize that in adding the amplitudes and then squaring, we are assuming that there is coherence between the states. If there were no coherence then the expression above for $|M|^2$ would be valid for any individual atom, but when averaged over the entire ensemble of atoms would give a simple exponential decay. The oscillation in the intensity given by (18.38) has been observed, and is referred to as quantum beating.

Further, if we measured the energy of the photons so that we distinguished between the decays $\Delta F = 0$, $\Delta F = +1$, and $\Delta F = -1$, then we would find simply that $|M|^2 \propto \exp(-t/\tau)$. Formally, the absence of quantum beats under these conditions would result from squaring and then adding the matrix elements for each of these decays, which are now distinguishable. Physically, in order to measure with some certainty whether the photon came from a $\Delta F = 0$ or a $\Delta F = +1$ transition, by measuring its energy precisely, by Heisenberg's uncertainty principle one would have to observe the photon for a time much greater than $1/A$. Such an observation would have the effect of washing out the cosine terms appearing in the squared matrix element (18.38).

Solution 9.6.   There are many approaches to each of the parts of this question. The solutions presented here are representative of our personal bags of physics tricks; the reader may find that her or his background makes other approaches easier.

a) A microwave oven has a typical dimension of about 35 cm. The wavelength of the microwave radiation should be, say, a quarter of this distance so that there will be a reasonable number of antinodes across the oven and food will be heated relatively evenly. A wavelength of 9 cm corresponds to a frequency of about 3 GHz.

b) In a fission bomb there are about $(30 \text{ kg})/(238 \text{ g}) = 126$ moles $= 7.6 \times 10^{25}$ atoms of uranium. A typical energy release in a fission process is 200 MeV. In an explosion, only a small portion of the uranium fissions before the material is blown apart, say ten percent. Then

$$E \approx 7.6 \times 10^{25} \times (200 \text{ MeV}) \times 10\% = 2 \times 10^{27} \text{ MeV}$$
$$= 2 \times 10^{14} \text{ J}. \qquad (18.39)$$

This is the equivalent of about 60 kilotons of TNT, which is about three times the yield of the bomb dropped on Hiroshima during World War II.

c) The energy depends on the relative velocity at impact. The earth's orbital velocity is about 30 km/s (since it traverses a circular route of radius 8 light-minutes every year). Take this as a typical relative velocity. We guess that the density of the meteorite is about that of ice, 1 g/cm^3, so

$$E \approx \frac{1}{2}mv^2 \approx \frac{1}{2} \left(\frac{4}{3}\pi \left(\frac{10 \text{ m}}{2} \right)^3 \right) \left(10^3 \frac{\text{kg}}{\text{m}^3} \right) \left(30 \times 10^3 \frac{\text{m}}{\text{s}} \right)^2$$
$$\approx 2 \times 10^{14} \text{ J}. \qquad (18.40)$$

This is the same as our answer to part (b)!

d) We recall that the sun's energy flux at the earth's surface (the solar constant) is $F_r = 0.14$ W cm^{-2}. The earth's orbit has radius $r \approx 8$ light minutes, and the radius of the sun is $R \approx 2$ light seconds, so the radiation flux at the surface of the sun is

$$F_R = F_r \left(\frac{r}{R} \right)^2 = 8 \times 10^3 \text{ W cm}^{-2}. \qquad (18.41)$$

The sun is very nearly a blackbody, so to find the temperature we use the Stefan-Boltzmann law, $F = \sigma_B T^4$, where $\sigma_B \approx 6 \times 10^{-12}$ W/cm^{2}K^4. Then $T \approx 6000$ K, which is about right.

e) We assume the filament is radiating as a blackbody, and use Wien's displacement law, $T\lambda_{max} = 0.29$ cm K. We need to estimate the wavelength at which a light bulb spectrum peaks. Efficiency requires that this should be at least near the optical band, to keep from wasting a lot of power in the UV (if T is too high) or the infrared (if T is too low). So we will assume, somewhat arbitrarily, that the peak is on the boundary between the optical and infrared bands, $\lambda_{max} \approx 1$ micron (recall that 1 micron = 10^{-6}m). Then $T \approx 2900$ K. (The melting point of tungsten is about 3700 K, which provides a useful upper limit.)

f) The speed of sound in a gas is

$$v = \left(\frac{\gamma P}{\rho}\right)^{1/2}, \tag{18.42}$$

where P and ρ are the gas pressure and density, respectively, and $\gamma = C_P/C_V$. Using the known speed of sound in air, 330 m/s, and the fact that air is largely molecular nitrogen, we can find the speed of sound in He. We will neglect the differences in γ between the two gases. For a given pressure, we have $v \sim 1/\sqrt{m}$ where m is the mass of a gas molecule, so

$$v(\text{He}) \approx v(\text{air})\sqrt{\frac{m(\text{N}_2)}{m(\text{He})}} \approx (330 \text{ m/s})\sqrt{\frac{28}{4}} \approx 870 \text{ m/s}. \tag{18.43}$$

The experimental result is about 965 m/s.

g) We guess that the average human has about four liters of blood, half of which is in the capillaries at any given time. Then the total volume of capillaries is two liters, or $V \sim 2 \times 10^{-3}$ m^3. A blood cell is about a micron across, and we recall from high school health films that capillaries are not much wider than the cells they transport, so we guess that a typical capillary is about $d \sim 3$ microns in diameter. Thus the total length of capillaries is about

$$L \approx \frac{V}{d^2} \approx 2 \times 10^8 \text{ m}. \tag{18.44}$$

(This is half the distance to the moon....)

h) In the temperature range where $c_V = 3k/2$, the only excited energy modes of the molecular hydrogen correspond to the three spatial translations. The transition to $c_V = 5k/2$ occurs when the temperature is high enough to excite the rotational modes of the molecule (recalling that the axial symmetry of H_2 implies that it has only two rotational modes). An additional complication is that the states with odd values of J are forbidden by the mirror symmetry of the molecule, so the lowest rotational state occurs at $J = 2$. The rotational energy of the molecule with $J = 2$ is

$$E_R \approx \frac{J(J+1)\hbar^2}{2I} \approx \frac{6\hbar^2}{2Ma^2}, \tag{18.45}$$

where I is the moment of inertia of the system, M is the total mass, and a is the separation between the hydrogen atoms. Using the estimates $a \approx 1$ Å and $M \approx 2 \times 940$ MeV, as well as the handy conversions $\hbar c = 197$ MeV-fm and $kT \approx 1/40$ eV at room temperature (300 K), we find $T \approx 74$ K. The measured value is really twice this.

Solution 9.7. a) We simply write down the ground-state electronic configurations, filling the states in the usual sequence of

$$1s\, 2s\, 2p\, 3s\, 3p\, 4s\, \ldots. \tag{18.46}$$

This gives us the following:

Element	Configuration
Li($Z = 3$)	$1s^2 2s^1$
B($Z = 5$)	$1s^2 2s^2 2p^1$
N($Z = 7$)	$1s^2 2s^2 2p^3$
Na($Z = 11$)	$1s^2 2s^2 2p^6 3s^1$
K($Z = 19$)	$1s^2 2s^2 2p^6 3s^2 3p^6 4s^1$.

(18.47)

(Note that we use lower-case letters to denote quantum numbers for single electrons, and capital letters to signify quantum numbers for the whole atom.) To find the ground-state L, S, and J, we recall Hund's rules:

1. maximize S,
2. maximize L (*after* maximizing S), and
3. for a shell labeled by n,
   · *minimize J* if the shell is less than half full, or
   · *maximize J* if the shell is more than half full.

All of the elements listed except nitrogen have only one valence electron, so $S = 1/2$. The valence electrons of Li, Na, and K are each in s-states, so that $L = 0$ for each of these elements, and thus $J = 1/2$. For boron, we have one electron in a p-state, so $L = 1$. We minimize J because the $n = 2$ shell has only three out of eight possible states occupied, so $J = 1/2$.

Nitrogen has three valence electrons. We apply the first rule above to find $S = 3/2$ for N. Because $S = 3/2$, we know that all three valence electrons have the same spin quantum number m_s. (That is, all three spins are parallel to one another.) The electrons already have identical s and l quantum numbers, so their values of m_l must all be distinct: $m_{l_1} = +1$, $m_{l_2} = 0$, and $m_{l_3} = -1$. Since the only possible state is the one in which $M_L = 0$, it must be that $L = 0$ and thus $J = 3/2$. To summarize:

$$
\begin{array}{lccc}
\text{Element} & S & L & J \\
\hline
\text{Li}(Z = 3) & 1/2 & 0 & 1/2 \\
\text{B}(Z = 5) & 1/2 & 1 & 1/2 \\
\text{N}(Z = 7) & 3/2 & 0 & 3/2 \\
\text{Na}(Z = 11) & 1/2 & 0 & 1/2 \\
\text{K}(Z = 19) & 1/2 & 0 & 1/2.
\end{array}
\tag{18.48}
$$

b) The lowest frequency line in the absorption spectrum corresponds to an electron in the $3s$ level absorbing a photon and being excited to a $3p$ state. The $3p$ state consists of two substates, $3p_{1/2}$ and $3p_{3/2}$, where the subscript is the J quantum number. The J states are split by the spin-orbit effect. The spin-orbit term of the hamiltonian arises because the electron sees a charged nucleus rotating around it, giving rise to a magnetic field, $\mathbf{B}$. The energy correction for this effect is given by $E_{so} \approx -\boldsymbol{\mu}_s \cdot \mathbf{B}$, where $\boldsymbol{\mu}_s = -g_s \mu_B \mathbf{S}/\hbar$ is the magnetic moment of the electron. The value of **B** is

$$
\mathbf{B} \approx \frac{1}{2}\left(\frac{\mathbf{v}}{c} \times \mathbf{E}\right) \approx \frac{1}{2}\left(\frac{\mathbf{v}}{c} \times \left(\frac{e}{r^2}\right)\hat{\mathbf{r}}\right)
$$
$$
\approx \frac{e}{2mcr^3}(m\mathbf{v} \times \mathbf{r}), \tag{18.49}
$$

where the factor of $1/2$ is from the Thomas effect, included here for completeness (see Jackson). (Also note that in using $\mathbf{E} \approx (e/r^2)\hat{\mathbf{r}}$ we are assuming that the effective charge of the shielded nucleus is roughly $+e$.) We recognize $\mathbf{L} = m(\mathbf{v} \times \mathbf{r})$, so we find that the spin-orbit energy is

$$
E_{so} = \frac{g_s \mu_B e}{2mc\hbar r^3}\mathbf{L} \cdot \mathbf{S} = \frac{g_s \mu_B^2}{\hbar^2 r^3}\mathbf{L} \cdot \mathbf{S}. \tag{18.50}
$$

Thus the energy difference between the two $2p$ states is proportional to $\langle r^{-3}\rangle$, so that $n = -3$.

c) We want to find the L, S, J, and M_J quantum numbers for the low-lying states. We will use spectroscopic notation in which the states are labeled as $^{2S+1}L_J$. The ground state is $4s^1$, or $^2S_{1/2}$ in spectroscopic notation. The lowest excited states have an electron in the $4p$ state, outside a filled core. Any one-electron state has $s = 1/2$ so that $S = 1/2$, and similarly $L = l = 1$ so that $J = 1/2$ or $3/2$. This means that the low-lying states are $^2P_{1/2}$ and $^2P_{3/2}$. The magnetic field breaks the degeneracy with respect to M_J.

Electric dipole ($E1$) transitions must obey the following rules:

1. $\Delta l = \pm 1$ (provided that the transition is of a single electron),
2. $\Delta S = 0$ (if L-S coupling is valid),
3. $\Delta L = 0, \pm 1$ but $L = 0 \not\to L = 0$ (if L-S coupling is valid),
4. $\Delta J = 0, \pm 1$ but $J = 0 \not\to J = 0$ (in general),
5. $\Delta M_J = 0, \pm 1$, but $M_J = 0 \not\to M_J = 0$ if $\Delta J = 0$ (in general).

These rules permit the transitions shown in Figure 18.5.

The energy change for a magnetic moment $\boldsymbol{\mu}$ in a field **B** is $\Delta E = -\langle \boldsymbol{\mu} \cdot \mathbf{B}\rangle$, or simply $\Delta E = -B\langle \mu_z\rangle$ if we take B to be in the $+\hat{\mathbf{z}}$-direction. Here, $\boldsymbol{\mu} = -(g_L \mu_B \mathbf{L} + g_S \mu_B \mathbf{S})$. For an electron, $g_S \approx 2$ and $g_L = 1$. We may rewrite $\boldsymbol{\mu}$ as

$$
-\frac{\boldsymbol{\mu}}{\mu_B} = \frac{1}{2\hbar}(g_L + g_S)(\mathbf{L} + \mathbf{S}) + \frac{1}{2\hbar}(g_L - g_S)(\mathbf{L} - \mathbf{S}). \tag{18.51}
$$

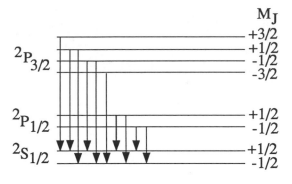

Figure 18.5.

We next use the expression for $\langle \mu_z \rangle$ from the projection theorem (which is also used in calculating nuclear Schmidt limits, Problem 7.2),

$$\langle \mu_z \rangle = \frac{\langle \boldsymbol{\mu} \cdot \mathbf{J} \rangle}{J(J+1)\hbar^2} \langle J_z \rangle. \tag{18.52}$$

The projection theorem is just the Wigner-Eckart theorem for diagonal matrix elements. With our expression for $\boldsymbol{\mu}$, equation (18.51), we find

$$\langle \mu_z \rangle = -\frac{\mu_B}{J(J+1)} \left[\frac{3}{2} J(J+1) - \frac{1}{2} L(L+1) + \frac{1}{2} S(S+1) \right] M_J. \tag{18.53}$$

We may thus write $\Delta E = g_J \mu_B M_J B$, where the Landé g-factor g_J is given by

$$g_J = \frac{3}{2} - \frac{L(L+1) - S(S+1)}{2J(J+1)}. \tag{18.54}$$

Thus, the splittings of the Zeeman levels in the $^2P_{3/2}$, $^2P_{1/2}$, and $^2S_{1/2}$ levels are governed by the Landé g-factors: $g_{3/2} = 4/3$, $g_{1/2} = 2/3$, and $g_{1/2} = 2$, respectively. (Of course, the $^2P_{3/2}$ and the $^2P_{1/2}$ levels are split by the spin-orbit effect of part (b).)

Solution 9.8.   a) For small velocities of the projectile, it is in the regime of low Reynolds number, in which viscous drag forces will dominate. The drag on a sphere is consequently given by Stokes's law, $F = 6\pi R v \eta$. We could just write this down, but it is more instructive to derive it (to within a numerical factor) from the basic equation of fluid mechanics, the Navier-Stokes equation:

$$\rho \left(\frac{\partial \mathbf{u}}{\partial t} + \mathbf{u} \cdot \boldsymbol{\nabla} \mathbf{u} \right) = -\boldsymbol{\nabla} p + \eta \boldsymbol{\nabla}^2 \mathbf{u} . \tag{18.55}$$

In this equation, $\mathbf{u}$ is the velocity field of the fluid, and p is the pressure. The two terms on the left-hand side together form the convective derivative of the fluid flow, and represent the acceleration experienced by a particle moving along with the fluid. On the right-hand side is a term for the pressure gradients which drive the motion of the fluid, and a term for the viscous damping of the flow. Mathematically, the effect of the spherical projectile on the fluid is to define boundary conditions on the fluid at the surface of the sphere, namely that the fluid must have zero velocity relative to the projectile at its surface (which is a standard condition when considering viscous fluids).

A lot of information can be extracted from the Navier-Stokes equation simply by using very crude approximations. First let us assume that the two acceleration terms on the left-hand side of (18.55) have the same magnitude. We can approximate $|\boldsymbol{\nabla} \mathbf{u}|$ by $\Delta u / \Delta x \approx v/R$, and in fact whenever we encounter a derivative $\boldsymbol{\nabla}$, we will replace it by $1/L$, where L is the appropriate length scale. In this problem, there is only one candidate for this length, the radius R of the projectile. Thus we make the estimate

$$\left| \rho \frac{\partial \mathbf{u}}{\partial t} \right| \approx |\rho \mathbf{u} \cdot \boldsymbol{\nabla} \mathbf{u}| \approx \frac{\rho v^2}{R} . \tag{18.56}$$

In a similar fashion we can estimate the viscous damping term to be of order

$$\left| \eta \boldsymbol{\nabla}^2 \mathbf{u} \right| \approx \frac{\eta v}{R^2} . \tag{18.57}$$

For arbitrarily small v, the acceleration terms will be negligible in comparison with the viscosity term. Therefore the viscosity term must balance against the pressure term,

$$|\boldsymbol{\nabla} p| \approx \frac{\Delta p}{R} \approx \frac{\eta v}{R^2} . \tag{18.58}$$

We interpret this as follows. The viscosity of the fluid gives rise to a pressure gradient across the projectile. The pressure will be greater on the front than on the back, and the pressure difference Δp is the source of the drag on the projectile. The magnitude of the drag force will be approximately

$$F \approx (\text{Area})\Delta p \approx \pi R^2 \frac{\eta v}{R} = \pi \eta v R, \tag{18.59}$$

which is a good estimate of Stokes's law, the exact result.

b) For high velocities, the acceleration terms in (18.55) will be far larger than the viscosity term, and we must balance them against the pressure term:

$$\frac{\Delta p}{R} \approx \frac{\rho v^2}{R}, \tag{18.60}$$

or $\Delta p \approx \rho v^2$. Thus the drag force in this regime is

$$F \approx (\text{Area})\Delta p \approx \rho v^2 R^2 . \tag{18.61}$$

Consider the work done by the force over a short distance,

$$\int F \, dx = \rho v^2 \, R^2 \Delta x . \tag{18.62}$$

This work looks like the kinetic energy of a volume $R^2 \Delta x$ of fluid moving at velocity v. We interpret this in terms of the projectile moving into regions where the fluid is at rest, and leaving in its wake a column of fluid which is eddying with circular velocity v (as shown in Figure 18.6). The projectile imparts kinetic energy to the fluid, and consequently must lose energy, and this gives rise to a retarding force.

We find the characteristic velocity marking the crossover between these two regimes by setting equal the two expressions for F: $\rho v^2 R^2 \approx \eta v R$. From this we find that the characteristic velocity is $v_c \approx \eta / \rho R$.

c) If the projectile starts off with velocity $v_0 \gg v_c$, then we must consider how far it will travel in each of the two regimes separately. While the projectile is still traveling rapidly, its equation of motion is

$$F \approx \rho v^2 R^2 \approx -\mu R^3 \frac{dv}{dt}, \tag{18.63}$$

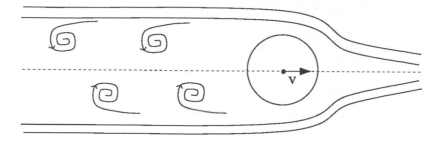

Figure 18.6.

where we have dropped numerical factors. We integrate once to get the velocity as a function of time,

$$v(t) \approx v_0 \left(1 + \frac{\rho v_0 t}{\mu R} \right)^{-1}, \qquad (18.64)$$

and once again to find the distance traveled,

$$x(t) \approx \frac{\mu R}{\rho} \ln \left(1 + \frac{\rho v_0 t}{\mu R} \right). \qquad (18.65)$$

Of course, this expression is only valid provided the projectile is still traveling at velocity $v(t) \gg v_c$. We use expression (18.64) for $v(t)$ to estimate the time t_c during which this approximation is valid:

$$v(t_c) = v_c \approx v_0 \left(1 + \frac{\rho v_0 t_c}{\mu R} \right)^{-1}, \qquad (18.66)$$

which we can solve to find $t_c \approx \mu R / \rho v_c$. We substitute into (18.65) to find the distance traveled in this time:

$$x \approx \frac{\mu R}{\rho} \ln \left(1 + \frac{v_0}{v_c} \right). \qquad (18.67)$$

The particle has not yet stopped completely, and we should now solve the equation of motion to see how far it travels in the regime of low velocities,

$$F \approx \eta R v \approx -\mu R^3 \frac{dv}{dt}. \qquad (18.68)$$

After integrating twice we find the projectile travels the further distance

$$\Delta x \approx \frac{\mu R}{\rho} . \tag{18.69}$$

However, since $v_0/v_c \gg 1$, this extra distance is negligible.

Solution 9.9. a) Each new signal photon requires the destruction of a pump photon, so that the overall number of photons is conserved, or

$$\frac{I_p(0)}{\omega_p} \approx \frac{I_s(z)}{\omega_s} + \frac{I_p(z)}{\omega_p} , \tag{18.70}$$

where we have used (on the LHS) the initial condition that $I_p(0) \gg I_s(0)$. Solving for $I_p(z)$ and substituting into the equation for the signal growth (9.2), we find that

$$\frac{1}{I_s(z)} \frac{d}{dz} I_s(z) = g \left[I_p(0) - \frac{\omega_p}{\omega_s} I_s(z) \right] . \tag{18.71}$$

This can be integrated and solved for $I_s(z)$ using the method of partial fractions. The result is

$$I_s(z) = \frac{\omega_s}{\omega_p} I_p(0) I_s(0) e^{I_p(0)gz} \left(\frac{\omega_s}{\omega_p} I_p(0) - I_s(0) + I_s(0) e^{I_p(0)gz} \right)^{-1} . \tag{18.72}$$

Taking the small signal limit $I_s(z) \ll I_p(z)$ allows us to ignore the terms in the denominator containing $I_s(0)$, yielding

$$I_s(z) \approx I_s(0) e^{g I_p(0) z} , \tag{18.73}$$

which shows the expected exponential gain in the signal before saturation is reached.

  b) The state $|i\rangle$ is virtual; the atom only stays in it a "short" time Δt. The uncertainty principle tells us that the energy of that state is not well-determined, i.e., it has a width given by $\Delta E \Delta t \approx h$. The energy of the virtual state $|i\rangle$ (and thus the energy of the outgoing signal photon) depends on the energy of the incident pump photon. The frequency of the transition from $|i\rangle$ to $|f\rangle$ thus depends on the frequency of the incident (nearly monochromatic) pump photon. This frequency is tunable.

c) We assume that the transitions are electric dipole. In that case, state $|i\rangle$ must have parity opposite that of states $|g\rangle$ and $|f\rangle$. In strontium, the outer two electrons are in a $(5s)^2$ configuration, so the parity of the ground state is positive. The state $|i\rangle$ must therefore have negative parity, which could result from the configuration $(5s)(5p)$. The final state must have positive parity, and a lower energy than $|i\rangle$. A suitable configuration is $(5s)(4d)$.

Solution 9.10. a) A hurricane is a region of low pressure, with a diameter on the order of one hundred kilometers, surrounded by circulating winds. Although the pressure difference between the eye of the hurricane and its outer edge is large (a sizable fraction of the total atmospheric pressure), the radius of the hurricane is so large that the pressure *gradient* is very small. Therefore, we have to take into account other small forces, in particular those arising from the rotation of the earth. The surrounding air would like to flow radially inward to equalize the pressure difference, but it is deflected by the Coriolis force $\mathbf{F} = -2m\omega \times \mathbf{v}$. We can see that this (pseudo-) force arises from simple considerations of angular momentum. Consider, for example, air flowing in from the south. As it moves north, its angular momentum about the earth's axis has to be conserved, but its perpendicular distance from the earth's axis of rotation is decreasing, which means that its angular velocity must increase to compensate, and will be higher than the local angular velocity. As a result, an observer on the ground will see the air veer eastwards. Conversely, air from the north will veer west. It is simple to calculate the exact forces involved, however we just need to note that the net effect is to set up a counter-clockwise rotation (as viewed from above) about the eye of the hurricane. (Of course, in the southern hemisphere the rotation would be clockwise.)

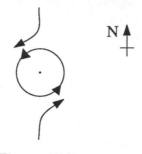

Figure 18.7.

b) Let us guess that the radius of the hurricane is about 50 km. Air pressure will vary from a minimum of 86% of normal atmospheric pressure at the center, to around 100% at the outer radius. We will approximate the pressure gradient as a constant, with value

$$\frac{dp}{dr} \approx \frac{0.14 \times (10^6 \text{ dyne/cm}^2)}{50 \times 10^3 \times (100 \text{ cm})} \approx 2.8 \times 10^{-2} \text{ dyne/cm}^3 . \qquad (18.74)$$

Now consider a packet of air of dimensions $L \times L \times L$, at a distance r from the center. It is circling the eye at velocity v and if we now ignore the Coriolis force (this only introduces a small error), the centripetal force required is provided by the pressure difference across the faces of the packet:

$$\Delta p \times (\text{Area}) = \left(\frac{dp}{dr} \times L\right) \times L^2 \approx -\rho L^3 \frac{v^2}{r} . \qquad (18.75)$$

Therefore the velocity is given by

$$v^2 \approx -\frac{r}{\rho}\frac{dp}{dr} . \qquad (18.76)$$

We see that at the eye of the hurricane the wind velocity is zero, as we would expect. It would also appear at first sight that the velocity increases without limit as r increases, but of course this is not the case as the pressure gradient vanishes sufficiently quickly.

Now we can calculate the wind speed near the eye, say at a radius of 10 km. At this radius we can assume that the pressure gradient is approximately the value calculated in equation (18.74). The corresponding velocity is

$$v \approx 5 \times 10^3 \text{ cm s}^{-1} \approx 110 \text{ mph}, \qquad (18.77)$$

which is a very reasonable number.

c) If we consider the pressure, temperature and density of air all as functions of height above sea level, the situation would be quite complicated. However, it is a good approximation to assume that the pressure is a constant, as it varies with height much more slowly than the other quantities.

Consider now a packet of air at height h, temperature T_0 and mass m, which rises adiabatically a distance Δh which is not necessarily infinitesimal. In the process it gains an amount $mg\,\Delta h$ in gravitational potential energy, and this gain in potential must come at the expense of its thermal energy, and consequently the gas must cool. Since we are dealing with a fixed mass of air at constant pressure, we will use c_p rather than c_v and write $mg\,\Delta h = -mc_p\,\Delta T$, or

$$\frac{\Delta T}{\Delta h} = \frac{-g}{c_p} = -9.8°\text{C km}^{-1}. \qquad (18.78)$$

This temperature gradient is known as the adiabatic lapse rate. Now suppose that the temperature of the surrounding air at height $h + \Delta h$ is higher than $T_0 - \Delta T$. The surrounding air will then be less dense than the packet, and the packet will sink back down. If, however, the surrounding air is cooler than $T_0 - \Delta T$, it will be denser and the packet will experience buoyancy and want to continue rising even faster. In fact, the temperature gradient found above is precisely the limiting condition for stability against buoyancy. If the temperature drops more rapidly than this, then the convection currents that arise will carry thermal energy and reduce the gradient until it reaches its limiting value. We would expect this to be the situation on a summer's day, when the air in the morning starts off cold, but is warmed through contact with the ground which is heated by the sun. Often the conditions will be right for "thermals," long columns of warm air rising at rates of a few hundred meters per minute, which glider pilots and flocks of birds use to gain height.

Note that there is no dynamical reason why the temperature cannot fall *less* rapidly than the limiting value, and indeed under certain conditions the temperature can actually *increase* with height. This is known as a temperature inversion, and can result in polluted city air being trapped close to the ground.

d) We assumed previously that the energy for the upward motion of the air came from its thermal energy. However, if the air is moist (in particular, if it is saturated as in a hurricane full of water that has evaporated from the sea), we can also derive energy to power the upward motion from the latent heat released through condensation of the water vapor, which will fall as rain. This will result in a smaller thermal gradient than that calculated above, but the actual value depends on more information than is given in the question, and is harder to calculate.

Bibliography

Ashcroft, N. W. and N. D. Mermin, *Solid State Physics*, Saunders, Philadelphia (1976).

Bahcall, J. N., *Neutrino Astrophysics*, Cambridge University Press (1989).

Boas, M. L., *Mathematical Methods in the Physical Sciences*, 2nd edition, Wiley, New York (1983).

Burcham, W. E., *Elements of Nuclear Physics*, Longman, Harlow (1979).

Cohen-Tannoudji, C., B. Diu, and F. Laloë, *Quantum Mechanics*, volume II, translated by S. R. Hemley, N. Ostrowsky, and D. Ostrowsky, Wiley, New York (1977).

Cottingham, W. N. and D. A. Greenwood, *An Introduction to Nuclear Physics*, Cambridge University Press (1986).

Frauenfelder, H. and E. M. Henley, *Subatomic Physics*, Prentice-Hall, Englewood Cliffs, New Jersey (1974).

Goldstein, H., *Classical Mechanics*, 2nd edition, Addison-Wesley, Reading, Massachusetts (1980).

Halzen, F. and A. D. Martin, *Quarks and Leptons: An Introductory Course in Modern Particle Physics*, Wiley, New York (1984).

Hecht, E. and A. Zajac, *Optics*, Addison-Wesley, Reading, Massachusetts (1974).

Hornyak, W. F., *Nuclear Structure*, Academic Press, New York (1975).

Huang, K., *Statistical Mechanics*, 2nd edition, Wiley, New York (1987).

Jackson, J. D., *Classical Electrodynamics*, 2nd edition, Wiley, New York (1975).

Jordan, T. F., Journal of Mathematical Physics **28**(8), 1759 (1987).

Kolb, E. W. and M. S. Turner, *The Early Universe*, Addison-Wesley, Redwood City, California (1990).

Landau, L. D. and E. M. Lifshitz, *Mechanics*, 3rd English edition, translated by J. B. Sykes and J. S. Bell, Pergamon, Oxford (1976).

Lightman, A. P., W. H. Press, R. H. Price, and S. A. Teukolsky, *Problem Book in Relativity and Gravitation*, Princeton University Press (1975).

Messiah, A., *Quantum Mechanics*, volume I, translated by G. M. Temmer, Wiley, New York (1958).

Perkins, D. H., *Introduction to High Energy Physics*, 3rd edition, Addison-Wesley, Menlo Park, California (1987).

Rose-Innes, A. C. and E. H. Rhoderick, *Introduction to Superconductivity*, 2nd edition, Pergamon, Oxford (1978).

Sakurai, J. J., *Advanced Quantum Mechanics*, Addison-Wesley, Redwood City, California (1967).

Schutz, B. F., *A First Course in General Relativity*, Cambridge University Press (1985).

Tinkham, M. *Introduction to Superconductivity*, McGraw-Hill, New York (1975).

Ziman, J. M., *Principles of the Theory of Solids*, 2nd edition, Cambridge University Press (1972).

Index